Probing the Past with Data Analytics and AI

Probing the Past with Data Analytics and AI

Marc Thuillard
Latena, Switzerland

W⬦ World Scientific

NEW JERSEY · LONDON · SINGAPORE · GENEVA · BEIJING · SHANGHAI · TAIPEI · CHENNAI

Published by

World Scientific Publishing Co. Pte. Ltd.

5 Toh Tuck Link, Singapore 596224

USA office: 27 Warren Street, Suite 401-402, Hackensack, NJ 07601

UK office: 57 Shelton Street, Covent Garden, London WC2H 9HE

Library of Congress Cataloging-in-Publication Data
Names: Thuillard, Marc author
Title: Probing the past with data analytics and AI / Marc Thuillard, Latena, Switzerland.
Other titles: Probing the past with data analytics and artificial intelligence
Description: New Jersey : World Scientific, [2025] | Includes bibliographical references and index.
Identifiers: LCCN 2025001835 | ISBN 9789819807369 hardcover |
   ISBN 9789819807376 ebook for institutions | ISBN 9789819807383 ebook for individuals
Subjects: LCSH: Artificial intelligence
Classification: LCC Q335 .T57 2025 | DDC 001.30285/63--dc23/eng/20250310
LC record available at https://lccn.loc.gov/2025001835

British Library Cataloguing-in-Publication Data
A catalogue record for this book is available from the British Library.

For any available supplementary material, please visit
https://www.worldscientific.com/worldscibooks/10.1142/14163#t=suppl

Desk Editors: Soundararajan Raghuraman/Steven Patt

Typeset by Stallion Press
Email: enquiries@stallionpress.com

To Claudia, Estelle, Xavier, and my parents

Foreword

The 2024 Nobel Prizes in Physics and Chemistry were attributed to Artificial Intelligence (AI) pioneers. The prizes indicate AI's expected impact on our lives. Thinkers worry that AI may dehumanize humans, which may be the case if machine judgment is given more weight in decision-making than humans. Another fear is that society and science may be contaminated by the "hallucinations" of AI, leading to wrong decisions and fake science!

Many researchers find verifying groundbreaking results obtained by AI and data analytics in their research field hugely challenging, which is a handicap for their acceptance. Researchers in the field of Humanities are confronted with AI results that sometimes contradict their knowledge. They are often in the frustrating situation of believing or rejecting AI results based on their experience. This book addresses these concerns by providing researchers with tools and knowledge to evaluate AI results critically.

This book provides a comprehensive introduction to AI and data analytics methods. The various concepts are introduced as needed throughout the chapters, building a solid foundation for understanding even the most challenging topics. For those seeking a deeper understanding, the book includes optional mathematical sections that explain key concepts further. The aim is to enable readers to confidently explore specialized articles on AI applications, ask critical questions, and assess the validity and trustworthiness of AI-generated results in digital humanities.

The general research community is highly fascinated with AI but also mistrusts many of its results. The number of disruptive AI applications in basic science and the humanities is still limited. Still, some disruptive applications provide important new results that would have been difficult to achieve otherwise. The soft unrolling and ink detection of the Herculaneum papyri belong to that category. The transcription of old handwritten Latin books is another example.

In many applications, AI does not reach a human level. Despite that, AI may offer new perspectives or highlight previously overlooked aspects. Some applications are impressive, like the digital restoration of murals, even if they do not achieve the same level of expertise as modern specialists.

Despite these limitations, AI can still play a valuable role in research by supporting tasks that require intensive or repetitive work, like classifying thousands of images. For example, in archaeology, AI can help classify pottery fragments based on their style and decoration. The critical evaluation of AI results, a central theme of this book, remains crucial in all applications.

AI is not limited to deep learning, and this book introduces classification methods based on phylogenetic trees and networks that have found many applications. We introduce simple methods to validate classification results. Their simplicity and broadness permit a researcher not familiar with the details of the methods, provided the data are available, to verify the results or at least to form his own opinion on the result pertinence. Simple validation is especially valuable as some results are sometimes controversial despite being validated by global approaches. As one says, the devil lies in the details!

The book intentionally covers a wide range of applications and methods. We examine AI applications in areas like prehistory, paleontology, historical linguistics, and comparative mythology and touch on more contemporary topics, such as the development of skateboards. Additionally, studying galaxies connects us to both past and present, depending on whether we focus on the light we receive or the moment it is emitted. Why is this a deliberate choice to deal with a large application spectrum?

A narrow focus can yield profound insights but also risk overlooking interdisciplinary connections. The broad scope allows for interconnecting the diverse applications of AI in probing the past, highlighting advancements, potential, and challenges across various domains. Depending on the field, the achievements of AI are already quite impressive or limited to feasibility studies. We hope the broad approach will permit knowledge transfer between domains and that some presented applications will trigger applications in other fields.

We carefully selected referenced papers, which was difficult considering improvements are frequently incremental. A large coverage of different topics automatically excludes very interesting work. Articles that deeply influenced applications were preferred. This choice bears the danger of not doing justice to some important articles. Therefore, we cite many review articles that allow readers to explore specialized topics further.

AI is unusual in the sense that many groundbreaking articles never got reviewed. The field advances are so rapid that the delay caused by the review made many researchers let the community judge the quality of their work. An article in an archive may have thousands of citations a few months after being put online, and the article may never be submitted to a journal despite its influence in AI. The book considers this special situation, and we make the reader aware that not all references have been peer-reviewed.

The book is divided into six chapters, with an additional part including three contributions on specific applications. The six chapters provide a broad overview of the most important developments in AI and data analytics, focusing on applications outside of biology. More generally, each chapter presents a main part presenting the different topics so that someone without a deep mathematical background can understand the main ideas, applications, and challenges. Complementary sections provide some mathematical insights. The text is constructed so readers may read or skip those sections depending on their interests.

The first chapter briefly introduces the most important classical classification techniques. The choice was set on methods that find broad applications in data analytics and AI further in the book. In this chapter, as in the rest of this book, we follow the strategy to explain and illustrate the main ideas behind the most advanced AI with minimal use of equations. The reader is furnished with many good references to enquire further.

The second chapter is on a family of Neural Networks described by the acronym "CNN", which stands for convolutional neural networks. For years, neural networks steadily developed and found applications in different fields, some described in my book *Wavelets in Soft Computing*. Despite many applications, the real breakthrough in artificial intelligence came with the development of CNN and the realization of the capabilities of large neural networks using a massive amount of data for learning.

The chapter presents the main ideas that made that development possible. The chapter discusses the resulting advances in the classification field. CNNs have sparked new ideas, enabling novel applications. This chapter describes some of the opportunities and challenges of automatically classifying artifacts or fossils, the successes of CNN in archaeological prospection and surveillance, and their potential for digital restoration using generative AI. The final section complements the presentation by explaining the main algorithmic ideas behind CNN.

Chapter 3 is on the second major development in neural networks that led to Large Language Models (LLM), which are part of our daily lives nowadays. The chapter first explains the concept of word embedding: transforming a word into a series of numbers. The corresponding vectors characterize words in a dictionary. We explain the extension of the concept to the attention mechanism, the core component of LLM. The application sections present applications to ancient languages.

Understanding the past requires tracing the evolution of ideas, languages, and cultures. While phylogenetic studies on ancient DNA have revolutionized our understanding of the past, their application outside biology presents unique challenges and opportunities.

Chapter 4 introduces the different phylogenetic methods and presents some specialized historical linguistic and paleontology models that extend phylogenetic studies to new domains outside biology. These models are central to the success of many cross-cultural case studies.

The applications of phylogenetic networks are the focus of Chapter 5. One of the key features of phylogenetic networks is their ability to represent lateral transfers in biological studies. The concept of lateral transfer extends to non-biological characters. For instance, a lateral transfer may correspond to borrowing a word from an unrelated language. For instance, the English word "algebra" comes from Arabic. Beyond instances of borrowing, phylogenetic networks can also capture broader patterns of cultural diffusion. A phylogenetic network may also result from a diffusion process with multiple origins for non-biological studies. It is a crucial distinction, as most studies outside of biology still assume a single origin corresponding to the unique root of a tree. This interpretation removes one of the major conceptual problems with the phylogenetic interpretation of phylogenetic networks outside biology. The chapter presents some applications on myths and starlores for which a diffusion process from multiple origins best explains the data.

In the last decades, Bayesian analysis became a pillar of artificial intelligence, and the sixth chapter is dedicated to that topic. Bayesian approaches have the great strength of integrating prior knowledge and uncertainty in the equations. The basic equation behind Bayesian analysis is extremely simple, but its application in real applications is not trivial. It requires manipulating probability distributions and sampling large data sets. The chapter introduces the reader to the Bayesian approach and presents some of the most advanced applications. Some results in paleontology and on the geographic origin of the Indo-European language challenge some recent views, triggering critical evaluation of both the models and existing knowledge. The final section discusses some ongoing debates on the accuracy of modeling and the validation of results.

The final part is a journey through time using astronomy as its guiding thread. It demonstrates how data analytics can extract meaningful insights from diverse sources, including scientific data, historical records, and cultural narratives. Didier Fraix-Burnet's contribution takes us deep into the past, focusing on the classification of galaxies and other celestial objects. Susanne Hoffmann and Boshun Yang provide a case study in historical astronomy. Using a Western (Ptolemy) and a Chinese (Master Shi) catalog, the authors extract information on potential early sources, the transmission of knowledge, and the quality of the instruments. The final contribution by Julien d'Huy brings "Man" into the picture, showing how folktale motifs are intertwined and how network theory quantifies their proximity. Let us note that the approach can also be applied to star-lores. This part shows the diversity of approaches to probe the past at different timescales with actual and exciting research examples.

I sincerely thank the competent and helpful team at World Scientific, particularly Steven Patt and Yolande Koh, for their great assistance and efficient support. I warmly thank the contributors for their insightful work: Didier Fraix-Burnet, Susanne Hoffmann, Boshun Yang, and Julien d'Huy. The many exchanges with research colleagues across various fields have been very helpful, especially Jean-Loic Le Quellec, Yuri Berezkin, Jean-Paul Marchand, Marc Frincu, and my past working colleagues from whom I learned so much. I am particularly thankful for the insightful discussions and invaluable advice from Roslyn Frank. I am deeply grateful for the support of my family, especially my son, Xavier Thuillard, who critically read the manuscript. Of course, any remaining errors are entirely my own.

About the Author

Marc Thuillard is a physicist (EPFL, Lausanne). He received a master's in Mathematics from Denver University and a PhD in Physics in Bern. After a postdoc at Caltech, he joined the industry as a researcher. After heading over 20 years the global research at a large Swiss company specializing in actuators, sensors, and energy optimization, he became Head of IP. Marc has written several books and book chapters on signal processing. Marc Thuillard has written his first book on Artificial Intelligence in 2001 *Wavelets in Soft Computing* which led to many "real-world" applications and products. In the last 10 years, Marc Thuillard has developed data analytics methods that were applied to cultural astronomy, and comparative mythology, methods that are also applicable in other fields such as the study of ancient DNA, computational linguistics, paleontology, cultural studies or archaeology. Marc Thuillard is now working as an independent consultant and author.

Contents

Part 1

Classification Methods

Chapter 1

Classification Methods for Machine Leaning and Artificial Intelligence (AI)

1.1 Introduction

Classification is a fundamental task in machine learning. This chapter introduces standard classification approaches. The main division is between supervised and unsupervised classification methods. In a typical unsupervised classification task, data and possibly some measures of uncertainty are supplied to a classifier. This approach often involves techniques like clustering or dimensionality reduction to discover patterns in unlabeled data.

In supervised learning, class labels are associated with the data points. Supervised learning aims to predict the output for new, unlabeled input data accurately. The input data can be obtained from various sources, including online platforms and offline datasets, such as images, text, or physical data. For example, the goal might be classifying images as containing cats, dogs, or birds. In natural language processing, one may want to categorize text as expressing positive or negative sentiment. The class label may have a semantic meaning (cat, sad, Brahms, bad weather) or be arbitrary (class A and B). While arbitrary labels can be useful, assigning semantic meaning to classes often provides more valuable insights. Knowing that a movie belongs to the same category as one just watched and liked may be useful when choosing between future viewings. Attributing semantic meaning to a class is often helpful (i.e., action movies instead of class A). Semantic labeling is a time-consuming and expensive task. Some of the most successful classifiers, like the convolutional neural networks (CNNs) introduced in the next chapter, attribute classes to input after a supervised learning phase in which the system is provided with labeled data, with the labels generally having a semantic meaning. Much ongoing research is focused on reducing the amount of labeled data in the learning phase without quality loss (see the following chapters).

We do not discuss methods exhaustively, as the literature on the topic is already very extensive. The scope is limited to methods encountered in the next chapters, such as those used in advanced neural networks. Several other classification methods are introduced in the text as necessary.

1.2 Pre-processing

The increasing volume and data complexity pose significant challenges for data analysis. As the saying goes: "garbage in, garbage out"; bad data can only furnish bad results. Pre-processing the data is, therefore, often an important stage in data analysis (Çetin and Yıldız, 2022). Pre-processing can take multiple forms (cleaning, normalization, and data selection). An important aspect is that the researcher gets familiar with the data. In many instances, the careful examination of the data leads to the conclusion that they are not what we believe to be! Some questions must be answered in that phase: are the data unnecessarily noisy or biased by some measuring errors? Are the data representative of the data that we want to classify? Should the data be filtered? Are they outliers? This process often eliminates many potential problems. Many discrepancies between results obtained with AI and specialist's knowledge turned out to be, after close examination, related to data that were not good enough. In some cases, it took years to understand the non-adequacy of some data. Such an example is the classification of Indo-European languages based on a set of cognates, where much improvement could be achieved using another data set (Heggarty *et al.*, 2023).

1.3 Unsupervised Classification

Unsupervised classification methods are often embedded in machine learning algorithms as a pre-processing or post-processing data component. For instance, in pre-processing, they can be used for feature extraction to identify the most relevant attributes, while in post-processing, they can group the results into meaningful clusters.

Many classification algorithms perform optimally when the data is represented in a low-dimensional space. In data science, the dimension corresponds to the number of variables or features used to describe a data point. Thousands of dimensions can be attached to a person, a region, or a word. The dimension of an image corresponds, for instance, to its pixel number. However, visualizing or analyzing data with thousands of dimensions can be extremely challenging. Therefore, dimension reduction approaches are crucial to many data analytics approaches and essential to identifying or visualizing relevant data. Given the challenges of working with high-dimensional data, dimension reduction techniques are essential for effective unsupervised classification. The section below introduces some fundamental approaches to dimension reduction.

1.3.1 *Dimension reduction (PCA and MDS)*

Unsupervised learning algorithms search for patterns, structures, or relationships within the data without being provided semantic information. Most classifiers do not work well at high dimensions if the space is mostly empty. For that reason, the dimensionality of the data is generally reduced using an appropriate algorithm. In the case of an image, a filter may be used to reduce the dimension, a classic approach in CNNs, discussed in the next chapter. Another simple approach consists of compressing the data into a matrix. The matrix summarizes the pairwise differences or similarities between the inputs. The principal component analysis (PCA) is the most popular dimensionality reduction method. A covariance matrix summarizes the input data, which are then projected into a new coordinate system. The new axes maximize the variance of the projected data (see, for instance, Greenacre *et al.*, 2022). The first principal component (PC1) is the direction along which the data has the most variance. The second principal component (PC2) is orthogonal to PC1 and captures the second most variance, and so on. Dimensionality reduction is achieved by focusing on the first few principal components, which capture most of the data's variance. A two-dimensional (2D) representation of the data will use the two main components.

The PCA approach is powerful, but the method also has some limits. For instance, in 2D, PCA is not a good approach if the projected data are distributed along a U-shape curve. Methods like multi-dimensional scaling (MDS) (Torgerson, 1952) or *t*-SNE (Van der Maaten and

Hinton, 2008) are often preferred for data with complex distributions. MDS aims to preserve the topological relations or relative distance between points in a low-dimensional projection as much as possible. It focuses on preserving the pairwise dissimilarities or distances between data points. The *t*-SNE approach is not much different from MDS. It calculates pairwise similarities between data points in the high-dimensional space using a Gaussian distribution and searches for a projection such as points that were close in the high-dimensional space are also close in the low-dimensional space.

The quality of the analysis is often judged by analyzing how well data points in the projection cluster in some meaningful separated clusters. These methods are included in most data processing software toolboxes, such as MATLAB or based on R.

Once data are projected in low dimensions, a clustering algorithm that works well in low dimensions is implemented to classify the data. The *K*-means clustering is a popular approach with many variants.

1.3.2 *Clustering*

1.3.2.1 *K-means clustering*

The *k*-means clustering approach is a classical classification approach with applications in almost all fields (Ahmed *et al.*, 2020; Ikotun *et al.*, 2023). The basic idea of *K*-means clustering is to partition the dataset into a specified number (K) of clusters, where each data point belongs to the cluster with the nearest mean (centroid).

K-means clustering:
(1) Attribute each data point to the nearest centroid among the K clusters.
(2) Update the cluster centroids to be the mean of the data points attributed to each cluster.
(3) Repeat until the stopping criterion is reached and iterate using different initialization values, as the result may strongly depend on the initial choice of the centroids.

The stopping criterion may correspond to reaching maximum iterations or apply when the centroids are no longer moving significantly.

Figure 1.1 shows an example of high-dimensional data being reduced to two dimensions and then classified with a *k*-means clustering

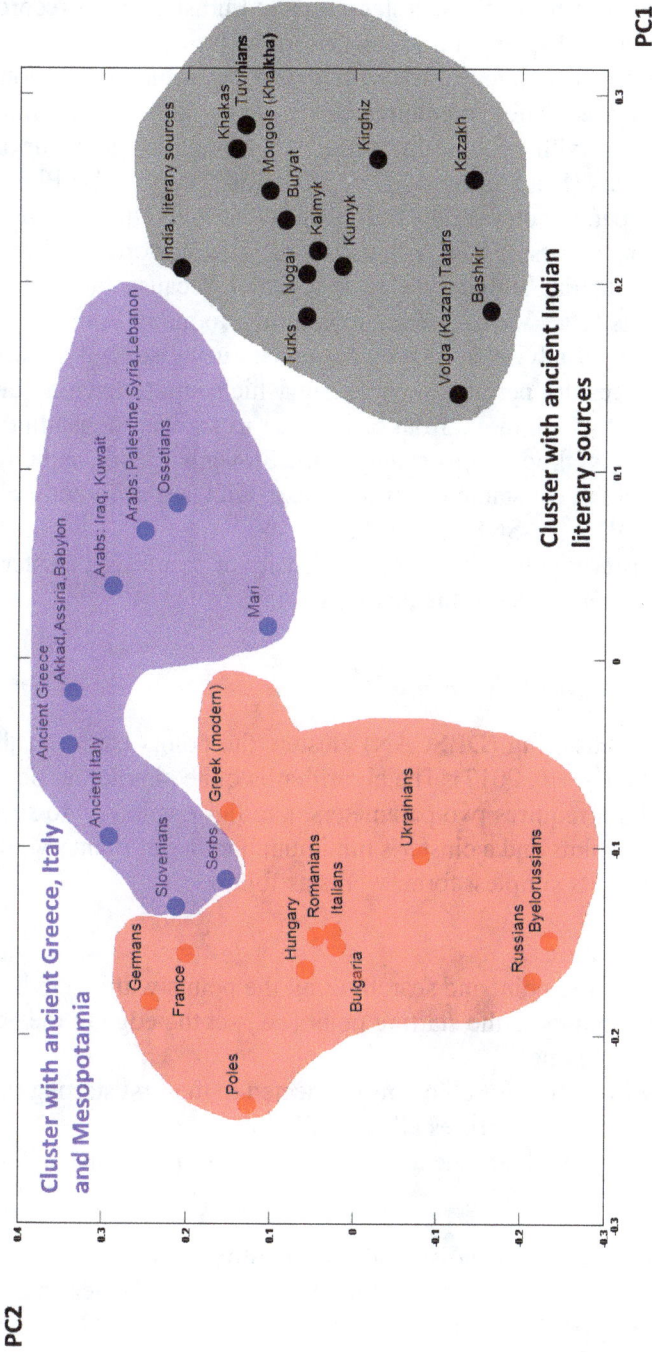

Fig. 1.1. Classification of the starlores motifs in Berezkin's database (Berezkin, 2015; Thuillard *et al.*, 2018) using a PCA approach to reduce the dimension of the data from sixty dimensions (binary state variables) to two dimensions. The data are classified in two dimensions with a *K*-means algorithm (*K* = 3). The colors of the data points correspond to the different clusters. The colored surface guides the eye.

algorithm. That example deals with star motifs in Eurasian myths recorded in many traditions. Chapter 5 further develops the topic.

The resulting clusters reveal interesting patterns in the distribution of starlore motifs across different cultures. For instance, some ancient Indian motifs are shared with present-day motifs in Mongolian and Turkish-speaking traditions (black cluster). They share motifs about the Pleiades and the Big Dipper. In contrast, the European cluster (red cluster) contains several common motifs related to agriculture. The starlores in Ancient Greece, Mesopotamia, and Italy cluster together because they contain unspecific motifs (e.g., Venus is female) common to many Eurasian star-lores. It is generally advisable to project the data using several methods. Comparing the results permits determining which inputs have a stable classification and which move from one cluster to another, depending on the classification method. In this example, the Slovenian, Serb, and Greek traditions do not have a stable classification, as they are differently clas-sified with an MDS or *t*-SNE approach.

Several clustering algorithms use the density of points as a clustering criterion. We present some of the methods below.

1.3.2.2 *Density-based clustering*

Density-Based Clustering (DBSCAN) clusters the points based on their density (Schubert *et al.*, 2017). The algorithm is quite simple but power-ful. The algorithm requires two parameters: a radius epsilon (ε) to search for neighboring points and a cluster's minimum number of points MinPts. The algorithm, in its simplest form, works as follows:

DBSCAN algorithm:
(1) Choose a starting point and search for all the points within a radius ε. Connect the points to the starting point (i.e., set the edge in the adja-cency matrix to one).
(2) Repeat the operation for all points connected to the first starting point until the algorithm identifies all cluster points.
(3) Repeat 1 and 2 with an unclassified point till all points are within a cluster.

DBSCAN is robust to noise and can identify clusters of arbitrary shapes and sizes. However, it may struggle with datasets where clusters have significantly different densities. Additionally, DBSCAN requires careful selection of the epsilon and MinPts parameters, which can affect the clustering results. DBSCAN is a simple method, but more advanced approaches based on density estimators are often preferred.

1.3.2.3 *Kernel density estimation*

The kernel algorithm for density estimation (kernel density estimation, KDE) is one of the preferred methods for detecting clusters. KDE is a non-parametric technique that estimates the probability density function (PDF) based on data points. It works by placing kernels, essentially smooth, symmetric functions, around each data point and summing them to create a continuous estimate of the PDF. The bandwidth of the kernel controls the smoothness of the density estimate. The complementary section 1.5 shows how KDE can be used to define cluster boundaries.

Kernel estimators and smoothing splines have numerous applications, for example, in archaeology (Ducke, 2015). KDE can be used to analyze the spatial distribution of artifacts within an archaeological site, revealing clusters that may indicate activity areas or areas of special significance. They can be used to estimate site occupations, surface material densities, Pb isotopic analysis for provenance studies, image processing, or the spatial arrangement of archaeological and prehistoric materials, a problem that already has a long history (Leroi-Gourhan and Laming-Emperaire, 1950; Lancelotti *et al.*, 2017).

The complementary section 1.5 provides a more detailed explanation of KDE. This section also discusses the broader context of regression analysis and how KDE can be used with other techniques. Professional software like ArcGIS implements KDE and other spatial analysis methods.

1.4 Supervised Learning and Classifiers

Again, we limit the discussion to the most popular methods, focusing on techniques that permit understanding some hybrid strategies that combine supervised learning with neural networks, as discussed in the following chapters.

1.4.1 *Support vector machine*

One popular supervised learning technique is the support vector machine (SVM). The algorithm divides the data space with a hyperplane, splitting the data into two classes and maximizing the distance of the closest points to the hyperplane (Vapnik, 2013). The method had many successful applications. A soft clustering variant allows some data points to be classified into both classes. Better classification results are generally obtained if data are transformed to a higher dimension, typically three dimensions, using a kernel transformation combining several variables, allowing for better separation of classes that are not separable at low dimensions. Several studies have shown the superiority of CNNs over SVM in real applications, for instance, the identification of cut marks and other bone surface modifications in an archaeological context (Byeon *et al.*, 2019).

1.4.2 *Linear and quadratic discriminant analysis*

Linear discriminant analysis (LDA) and quadratic discriminant analysis (QDA) are related. In 2D, LDA separates clusters with straight lines, while QDA uses more complex curves (quadratic functions) to define the boundaries between clusters. QDA is more flexible when dealing with complex cluster shapes but requires more data to work effectively.

Fisher laid the foundation for LDA. In his original paper, Fisher (1936) considered measurements of the characters of iris flowers and suggested a linear discriminant method to differentiate between two species. Given n characters, Fisher searches for the linear combination of characters:

$$S = \sum_i^n \lambda_i x_i$$

that best differentiates the two species. LDA is a supervised learning method, as the classifier learns with labeled examples to later distinguish between the classes using, for instance, a simple separation plane in three dimensions. Fisher suggests maximizing the ratio of the difference between the specific means to the sum of the standard deviation within species. Fisher's approach includes only the means and variances of the data.

The QDA includes the covariances and assumes that each class is normally distributed. The learning phase corresponds to computing the covariance matrix and the mean of the Gaussian function describing a class.

$$\text{Likelihood} \propto \frac{1}{\sqrt{\Sigma_k \pi_k}} \exp\left(-\frac{1}{2}\left(x - \mu_k\right)^T \Sigma_k^{-1}\left(x - \mu_k\right) \right) \quad (1.1)$$

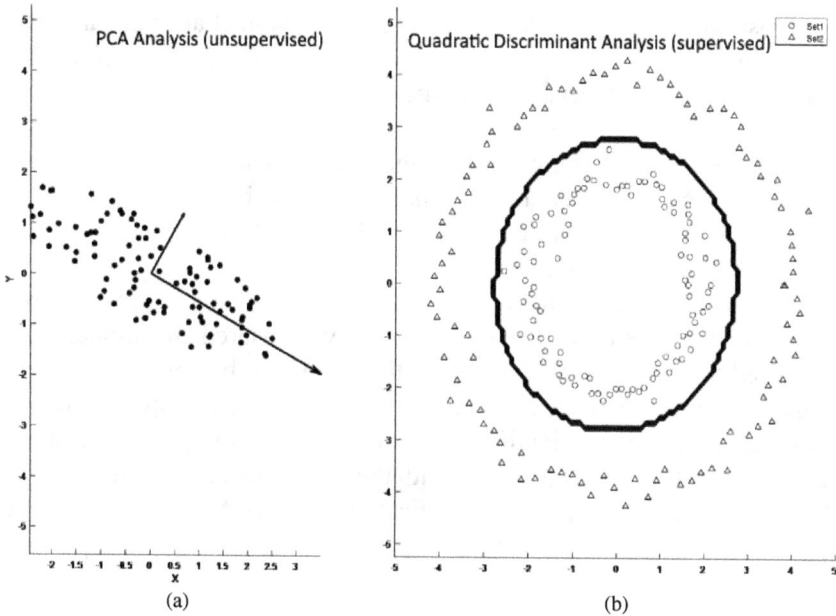

Fig. 1.2. (a) Example of a PCA analysis. The two main components correspond to the new axes. The data projected on the first component correspond to a dimension reduction from 2D to 1D. PCA is typically used on high-dimensional data. (b) The quadratic linear analysis separates the two sets of data. An LDA or a PCA would not be successful with such an example.

where T indicates the transpose, μ_k the average on the kth labeled subset, Σ_k the covariance matrix, and π_k the proportion of data with the kth label. The covariance of two variables, X and Y, is given by

$$\text{Cov}(X,Y) = \frac{1}{(N-1)} \sum_{i=1}^{N} (x_i - \mu_x)^2 \sum_{i=1}^{N} (y_i - \mu_y)^2 \qquad (1.2)$$

After learning, the algorithm attributes a class to each new input data by choosing the class with the highest likelihood for the input data. The resulting separating surface (or line in 2D) between the classes is quadratic.

Figure 1.2 illustrates the difference between LDA and QDA. In the example of Fig. 1.2(b), the two separating lines are represented only where they form a contour.

Most applications today combine several physical analysis and data processing techniques. Let us give two examples. Carvalho *et al.* (2020) analyzed ceramic objects from a site in west Amazonia using neutron activation analysis and thermoluminescence dating. A discriminant analysis classified the PCA components of the measured data and automatically separated the various clays into distinct clusters. This example shows that the input to a discriminant classifier can be some transformed variables, such as the ones after PCA.

Pobiner *et al.* (2023) identify butchery marks on hominin fossils from the early Pleistocene Turkana region. They measured 3D models and compared them with a database of individual teeth, butchery, and marks created through controlled experiments. They trained a quadratic discriminant estimator to identify cut marks. The resulting classifier automatically identifies butcher marks with the discriminant estimator (the 3D character of the data makes a classification with a CNN a less attractive option).

1.4.3 *Decision trees and random forests*

Decision trees and random forests are well-suited for handling complex datasets with nonlinear relationships. Classification and Regression Trees (CART), introduced by Breiman (2017), is a decision tree algorithm that recursively partitions the data. Each successive partition creates two new branches of a tree. The algorithm searches iteratively for the variable and the threshold value that best separates the different classes into two subsets of data. The best split maximizes or minimizes an index (entropy, Gini, information gain; Shaheen *et al.*, 2020). The data points with values above the threshold are attributed to one subset, and the remaining data points to another.

The random forest algorithm (Breiman, 2001) is an extension of CART. The algorithm builds multiple decision trees using the CART algorithm and combines their predictions. Each tree is built on a random subset of the data and variables at each split. This approach creates a diverse set of trees that capture different aspects of the data. The random forest algorithm predicts the class of a new data point by setting a vote. Each classification tree predicts a class, and the class with the most attributions (votes) is selected. The algorithm is quite good at integrating nonlinearities.

1.4.4 *Multiresolution fuzzy classifiers*

Multiresolution fuzzy classifiers offer a powerful approach to classification by combining wavelet analysis and fuzzy logic. Wavelets allow data analysis at different scales or resolutions, while fuzzy logic provides a framework for handling uncertainty and imprecision. This combination enables the development of classifiers that can capture complex patterns and relationships in the data.

Multiresolution classifiers can produce linguistically interpretable classification rules, such as "If *x* is large and . . . then 'bad weather'." This interpretability can be valuable in applications where understanding the reasoning behind the classification is important.

A more detailed discussion of multiresolution classifiers, including their theoretical foundations and practical applications, can be found in *Wavelets in Soft Computing* (Thuillard, 2023). The second chapter of this book discusses the concept of multiresolution in the context of CNNs in the complementary section.

1.5 Complement: Non-parametric Regression Techniques and Probability Density Functions

1.5.1 *Non-parametric regression*

Unlike traditional regression methods that assume a specific functional form (linear or polynomial), non-parametric regression uses kernel functions to determine the shape of the function. This flexibility makes it suitable for situations where the underlying relationship between variables is complex or unknown. The regression furnishes an estimation of the function in a p-dimensional space, with p corresponding to the kernel dimension.

The estimate is $\hat{f}(x)$ expressed as a weighted sum of translated and dilated kernels. The Watson–Nadaraya estimator is of the form:

$$\hat{f}(x) = \frac{\sum_{i=1}^{N} K\left(\frac{x_i - x}{h}\right) \cdot y_i}{\sum_{i}^{N} K\left(\frac{x_i - x}{h}\right)} \qquad (1.3)$$

with K centered on x_i and N, the number of points, and h the bandwidth. The data points are typically obtained experimentally and corrupted with noise. The Watson–Nadaraya estimator averages the data values around each point of interest. A large bandwidth results in a smooth estimate, while a small bandwidth captures more local fluctuations. Different kernels are used in applications, such as the Gaussian kernel. The Gaussian kernel is smooth, meaning the resulting PDF is smooth. The Gaussian kernel has infinite support, and all data points contribute to the estimation. The quadratic kernel is often referred to as the Epachnikov kernel. It minimizes the mean integrated square error when estimating a density function.

1.5.2 *Probability density function estimator*

The second related problem is estimating the PDF given many data points x_i. The PDF describes the likelihood of a continuous random variable taking a given value. Similarly to non-parametric regression, kernel density estimators (KDEs) implement kernel functions slightly differently. The KDE estimation uses typically an estimator of the form:

$$\widehat{f_h}(x) = \frac{1}{N} \sum_{i=1}^{N} K\left(\frac{x - x_i}{h}\right) \tag{1.4}$$

The KDE is normalized so that its integral equals one. Figure 1.3 illustrates the difference between kernel regression, which estimates a function, and KDE, which estimates a probability density.

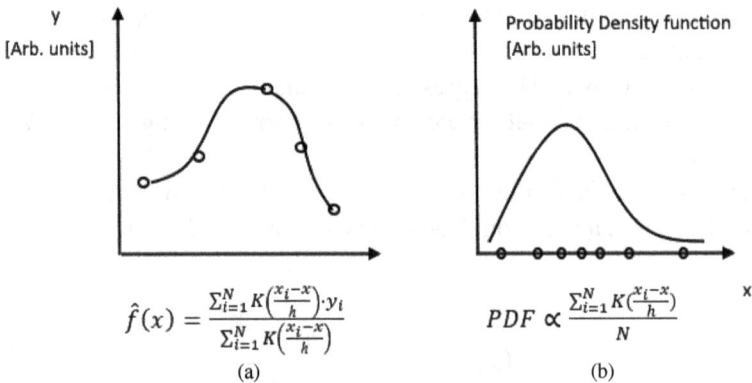

$$\hat{f}(x) = \frac{\sum_{i=1}^{N} K\left(\frac{x_i - x}{h}\right) \cdot y_i}{\sum_{i=1}^{N} K\left(\frac{x_i - x}{h}\right)}$$

(a)

$$PDF \propto \frac{\sum_{i=1}^{N} K\left(\frac{x_i - x}{h}\right)}{N}$$

(b)

Fig. 1.3.   Sketch of a kernel regression (a) and a KDE for a PDF (b). After Thuillard (2023).

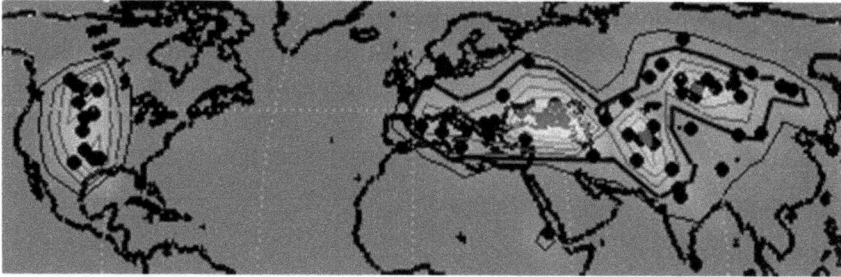

Fig. 1.4. KDE of the geographic distribution of a myth's motif, in which the seven stars of the Big Dipper are associated with seven men, using a Gaussian kernel. The approach permits data clustering the data into high-density regions. A cluster is defined by the constant density contour at a threshold value (thick line as an example). The number of clusters depends on the threshold value.

As an illustration, let us consider the geographic distribution of a specific motif in mythology, where the seven stars of the Big Dipper are associated with seven men. Figure 1.4 shows the locations (dots) of the traditions where this motif is present. We can estimate the underlying probability density of this motif using a Gaussian kernel. The probability is represented by shades of gray in Fig. 1.4, with white indicating a high probability. Areas with a higher concentration of traditions with the motif appear lighter in the figure. We can identify regions with a high motif concentration by drawing contour lines connecting points of identical probability density. For instance, the thick contour line in Fig. 1.4 highlights two major clusters: North America and Eurasia. If one uses a higher contour line, the Eurasian cluster breaks into three clusters. This method provides a rational and data-driven way to identify meaningful clusters and outliers.

By adjusting the bandwidth, we can control the level of detail and smoothness in the estimated probability density, revealing underlying patterns in the data. The choice of bandwidth is important as it may dramatically change the results. If the bandwidth is too large, the estimator may miss a superposition of two Gaussian. On the other hand, a bandwidth that is much too small may result in a density function consisting of single "peaks" around the data points. Silverman (2018) suggests rules for setting the bandwidth of univariate functions. The best bandwidth minimizes the mean squared error between the true and estimated densities. Under some conditions, bandwidth can be selected automatically based on the probability density. For regression problems, a better approach consists of dividing the data into training and validation sets and choosing the band-width that minimizes the mean integrated squared error of the validation test. Typically, the procedure is repeated many times in different training data. The method is often described as cross-validation.

References

Ahmed, M., Seraj, R., & Islam, S. M. S. (2020). The *k*-means algorithm: A comprehensive survey and performance evaluation, *Electronics*, 9(8), pp. 1295–1317.

Berezkin, Y. E. (2015). Folklore and mythology catalogue: Its lay-out and potential for research, *The Retrospective Methods Network*, (S10), pp. 58–70.

Breiman, L. (2001). Random forests, *Machine Learning*, 45, pp. 5–32.

Breiman, L. (2017). *Classification and Regression Trees* (Routledge, Milton Park UK).

Byeon, W., Dominguez-Rodrigo, M., Arampatzis, G., Baquedano, E., Yravedra, J., Maté-González, M. A., & Koumoutsakos, P. (2019). Automated identification and deep classification of cut marks on bones and its paleoanthropological implications, *Journal of Computational Science*, 32, pp. 36–43.

Carvalho, P. R., Munita, C. S., Neves, E. G., & Zimpel, C. A. (2020). A preliminary assessment of the provenance of ancient pottery through instrumental neutron activation analysis at the Monte Castelo site, Rondônia, Brazil, *Journal of Radioanalytical and Nuclear Chemistry*, 324, pp. 1053–1058.

Çetin, V., & Yıldız, O. (2022). A comprehensive review on data preprocessing techniques in data analysis, *Pamukkale Üniversitesi Mühendislik Bilimleri Dergisi*, 28(2), pp. 299–312.

Ducke, B. (2015). Chapter 18, Spatial cluster detection in archaeology: Current theory and practice, Barceló, J.A. and I. Bogdanović, (eds.), *Mathematics and Archaeology*, (CRC Press, Boca Raton) pp. 352–368.

Fisher, R. A. (1936). The use of multiple measurements in taxonomic problems, *Annals Eugenics*, 7(2), pp. 179–188.

Greenacre, M., Groenen, P. J., Hastie, T., d'Enza, A. I., Markos, A., & Tuzhilina, E. (2022). Principal component analysis, *Nature Reviews Methods Primers*, 2(1), p. 100.

Heggarty, P., Anderson, C., Scarborough, M., King, B., Bouckaert, R., Jocz, L., … & Gray, R. D. (2023). Language trees with sampled ancestors support a hybrid model for the origin of Indo-European languages, *Science*, 381(6656), p. eabg0818.

Ikotun, A. M., Ezugwu, A. E., Abualigah, L., Abuhaija, B., & Heming, J. (2023). *K*-means clustering algorithms: A comprehensive review, variants analysis, and advances in the era of big data, *Information Sciences*, 622, pp. 178–210.

Lancelotti, C., Negre Pérez, J., Alcaina-Mateos, J., & Carrer, F. (2017). Intra-site spatial analysis in ethnoarchaeology, *Environmental Archaeology*, 22(4), pp. 354–364.

Leroi-Gourhan, A., & Laming-Emperaire, A. (1950). *Les Fouilles Préhistoriques: Technique et Méthodes* (in French).

Pobiner, B., Pante, M., & Keevil, T. (2023). Early Pleistocene cut marked hominin fossil from Koobi Fora, Kenya, *Scientific Reports*, 13(1), p. 9896.

Shaheen, M., Zafar, T., & Ali Khan, S. (2020). Decision tree classification: Ranking journals using IGIDI, *Journal of Information Science*, 46(3), pp. 325–339.

Schubert, E., Sander, J., Ester, M., Kriegel, H. P., & Xu, X. (2017). DBSCAN revisited, revisited: Why and how you should (still) use DBSCAN, *ACM Transactions of Database Systems (TODS)*, 42(3), pp. 1–21.

Silverman, B. W. (2018). *Density Estimation for Statistics and Data Analysis* (Routledge, Milton Park UK).

Thuillard, M. (2023). *Wavelets in Soft Computing*, 2nd edn. (World Scientific), Singapore.

Thuillard, M., Le Quellec, J. L., d'Huy, J., & Berezkin, Y. (2018). A large-scale study of world myths, *Trames: A Journal of the Humanities and Social Sciences*, 22(4), pp. A1–A44.

Torgerson, W. S. (1952). Multidimensional scaling: I. Theory and method, *Psychometrika*, 17(4), pp. 401–419.

Van der Maaten, L., & Hinton, G. (2008). Visualizing data using t-SNE, *Journal of Machine Learning Research*, 9(11), pp. 2579–2605.

Vapnik, V. (2013). *The Nature of Statistical Learning Theory* (Springer Science & Business Media), Berlin/Heidelberg.

Chapter 2

Classification with Convolutional Neural Networks

2.1 Introduction

In 2019, the three pioneers (Yoshua Bengio, Yann Le Cun, and Geoffrey Hinton, who in 2024 received the Nobel price) were awarded the Turing Price, the most prestigious price in computer sciences, for their conceptual and engineering breakthroughs that made deep neural networks a central component of computational intelligence. After many years of very slow developments, their work came as a surprise and revolutionized the field of artificial intelligence by creating neural network architectures with impressive learning capabilities. They demonstrated the power of deep learning using large networks with multiple layers to learn and extract complex patterns from vast data. The expression "deep learning" refers to the many layers and feature maps of the neural networks. Deep learning is now an important subfield of machine learning (ML).

An aspect that contributed to the spectacular improvement in the performance of neural networks has been the availability of large databases for learning and the development of systems with huge numbers of parameters. Compared to most previous neural networks, convolutional neural networks (CNNs) have orders of magnitude more learned parameters whose values are determined during a training phase using labeled images (e.g., manually labeled images with cats, dogs, trains, and cars). A huge database of labeled images trains a large network, generally belonging to a broad class of networks called CNNs. Improvements in CNN have led to networks being capable of performing new tasks, such as image segmentation into multiple distinct regions based on certain features or characteristics or image generation. So, this chapter deals mostly with images!

CNN mostly relies on supervised learning. After a learning phase, the networks are often very good at classifying images. The technology has found many useful applications in medicine, with more than 700 AI medical getting FDA market-clearing of algorithms (stand: 2024), with more than two-thirds in radiology. CNNs are now implemented in many devices. The algorithms significantly impacted fast lung diagnosis during the COVID-19 epidemic (Kaul *et al.*, 2020; Sarvamangala and Kulkarni, 2022), demonstrating the real-world impact of CNN in medicine.

CNNs have applications in many fields, such as computer vision, remote sensing (Kattenborn *et al.*, 2021), autonomous driving (Yang *et al.*, 2024a, 2024b), fault analysis (Waziralilah *et al.*, 2019), weather prediction (Wang *et al.*, 2023) and applications in face recognition (Mehendale, 2020), a controversial topic. The success of CNN triggered a large effort in academia and industry that resulted in the invention of several innovative techniques that opened new application fields.

CNNs have great potential to support difficult tasks such as finding archaeological sites with remote sensing, automatically classifying the possible provenance and age of artifacts, or identifying pollen or fossils. This chapter introduces the most relevant deep learning techniques, focusing on the possibilities offered by the different approaches and furnishing selected examples in archaeology, epigraphy, and paleontology, presenting their strengths and weaknesses.

2.2 Convolutional Neural Networks

2.2.1 *CNN for classification and identification tasks*

A neuron is a fundamental building block of a neural network. It receives an input, processes it, and produces an output. A neural network consists of layers of parallel neurons, with the output layer being passed to the next layer after some transformation in which the output values are multiplied by some weights. A CNN consists of several layers using similar components. A final layer with a different architecture classifies the data. The system is trained with many labeled images (such as cars or cats). After training, the network typically recognizes an object in an image and outputs a category or class for that object (car, cat). A CNN performs very

Feature
maps

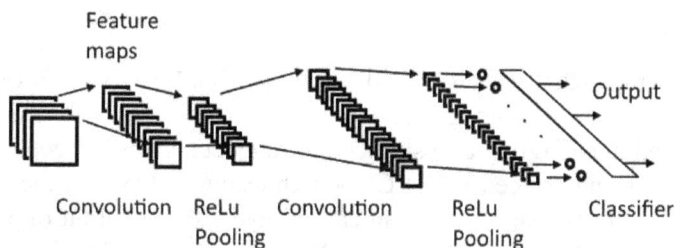

Convolution   ReLu   Convolution   ReLu   Classifier
           Pooling   Pooling

Fig. 2.1. Architecture of a typical CNN.

well on image processing tasks such as face recognition or object recognition in images.

Figure 2.1 shows a CNN architecture. CNNs use a cascade of simple linear convolution filters iteratively applied to neighboring pixels, downsampling, and nonlinear elements. The Complementary Section 2.6 introduces the main components of a CNN.

CNN's spectacular performances surprised the specialists in the field after many years in which neural networks could mostly solve simple tasks. It is quite astonishing that CNNs perform so well, considering the high dimension of the search space (the dimension of an image is the number of pixels). A possible explanation is that images have underlying relatively simple structures (see Section 2.7 for a more technical explanation).

The successive layers form complex representations, enabling the network to recognize objects even when rotated or dilated. Data augmentation techniques, such as rotating images during training or showing just an image section, can help CNN deal with objects of different sizes and positions.

CNNs learn by adjusting the weights and values of the filters, which are small matrices that detect specific features in the input data. The weights and filter values are optimized through gradient optimization during learning. The gradients are estimated through backpropagation, which consists of propagating the error signal back through the network's layers, allowing the network to learn and improve its performance. Despite impressive results, deep CNN architectures suffered from the vanishing gradient problem, which makes these networks difficult to train. It occurs because the gradient calculated during backpropagation becomes increasingly small as it propagates through layers, hindering proper learning. ResNet is a specialized CNN architecture that adds connections between layers to solve this issue.

2.2.2 *ResNet*

ResNet, which stands for Residual Networks, is a type of CNN designed to address the vanishing gradient problem. It uses "identity mapping", represented in a diagram as a shortcut, to reinject signals to deeper layers. Shortcuts are introduced in the CNN architecture. They add the original input signal, or the signal from an earlier layer, to the output of a deeper layer (Fig. 2.2). It ensures that information from earlier layers is not lost as it propagates through the network. This simple trick, proposed by He *et al.* (2016), was key to improving learning in deep CNNs and achieving better results. The name residual came from the suggestion in the original paper that the network learns residual mappings. Instead of trying to learn a function $H(x)$, the desired output, a residual block learns a residual function, $F(x) = H(x) - x$ with $F(x)$ the output of the neuron layers and x its input.

ResNet architectures typically skip two neuron layers with the identity mapping. Li *et al.* (2016) theoretically explained that skipping two layers contributes to their stability and learning capacity. This architecture is widespread due to its effectiveness in training deep networks.

2.2.3 *Region-based convolutional neural network*

Since the original work by LeCun, Bengio, Hinton, and coworkers in a series of papers (For the early history, see LeCun *et al.*, 2015), a few very clever architectures have opened the domain of application of

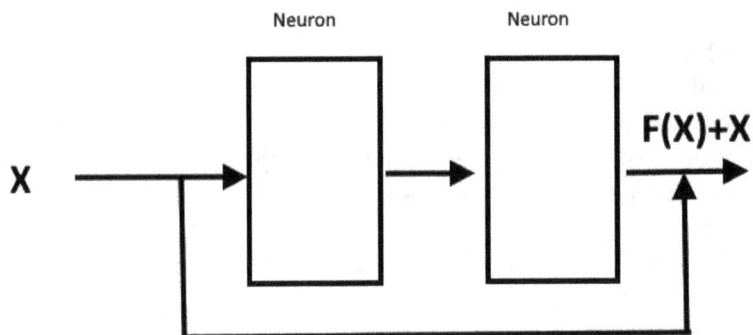

Fig. 2.2. ResNet is an advanced CNN architecture that facilitates learning. The output of a deep layer consists of adding the last layer processing and the signal at a higher level. The shortcut corresponds to identity mapping.

classification to complex images. While traditional CNNs focused on classifying an entire image (say, a cat or a dog), region-based CNN (R-CNN), introduced by Girshick *et al.* (2014), aimed to classify individual objects within an image and locate them precisely within a bounding box. A main improvement was achieved by assigning a class to each pixel (e.g., part of a cat) in the image (Long, 2015), enabling it to identify and delineate objects more accurately. The Facebook AI Research (FAIR) team further refined the approach with Mask R-CNN (He *et al.*, 2017), improving object detection, segmentation accuracy, and efficiency. The model is open-source and, therefore, is in great use.

Figure 2.3 illustrates neural networks' capabilities for image segmentation and object identification. The model recognizes each instance of an object and can name it, for example, a cucumber or a melon. Emerging applications in agriculture are numerous, such as automatically recognizing and localizing the origin of diseases or pests attacking fruits or determining their ripeness.

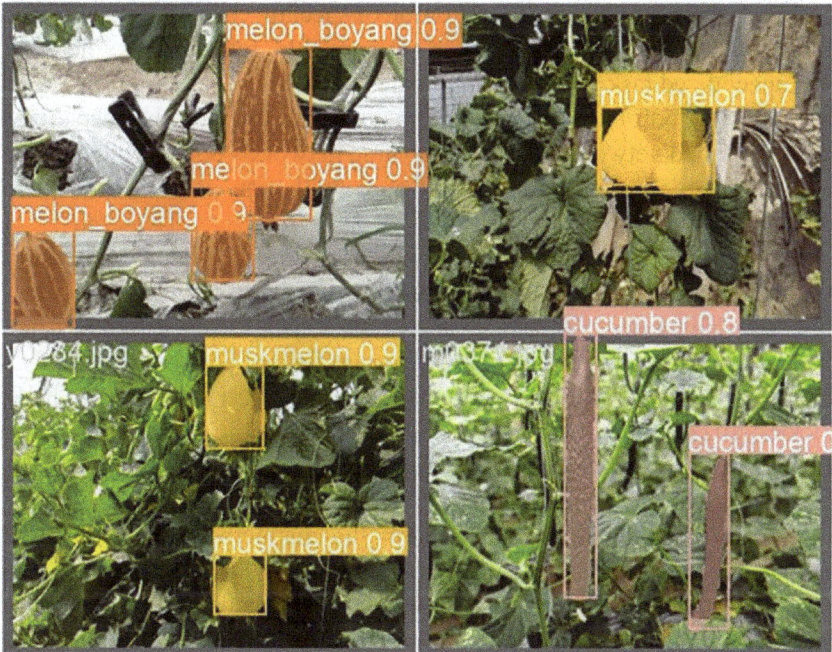

Fig. 2.3. Detection and classification of different objects with a CNN. From Lawal (2023), CC-BY license.

2.2.4 *Applications of CNN in archaeology*

2.2.4.1 *Prospection, surveys, and surveillance*

Applications of remote sensing using neural networks are well developed and meaningful, not only for prospection but also for the preservation and remote detection of damages to sites in wartime or regions that are difficult to access. R-CNN is one of the most popular networks proposed for segmentation. The code is available through the Detectron2 platform (Wu *et al.*, 2019). The transfer to archaeological applications was, therefore, quite rapid, as the architecture improves the quality of the interpretation of remote sensing data. The availability of enough labeled examples remains the main challenge of making a CNN learn the typical features of a structure. Large databases of labeled images are necessary to train very large models. Training of the complete models requires computing powers not available to most researchers. The best results are generally obtained by fine-tuning an open-source, very large model on some specialized data rather than training the model from scratch. This approach, also known as transfer learning, permits the integration of specialized knowledge (Oquab *et al.*, 2014). If the trained weights are available, retraining the network on a specialized task is a less complex task for which standard tools exist. Fine-tuning modifies some weight to be more efficient on the specialized task. The tuned network combines the large network's power with specific knowledge, for instance, images of ancient Iberian ceramics (Navarro *et al.*, 2021) to focus on those.

Large-scale archaeological prospection is a very difficult and tedious work that benefits from ML. The automatic discovery of interesting structures combining deep learning with remote sensing using LiDAR, satellites, or aerial remote sensing is an important field of applications of CNN and other neural networks (Jamil *et al.*, 2022; Caspari and Crespo, 2019; Bickler, 2021; Guyot *et al.*, 2021; Trier *et al.*, 2021). Caspari and Crespo (2019) apply CNN to analyze images of the Eurasian steppes, which is difficult due to the vastness and problematic access to the surveyed territories. They trained a CNN based on the distinctive shape of burial mounds to detect early Iron Age burials in satellite data. Berganzo-Besga *et al.* (2021) detected over ten thousand burial mounds in North-Western Iberia alone with a recall value of 64% (Recall = (True Positives) / (True Positives + False Negatives)), a quite good result.

R-CNN has been successfully applied to diverse tasks. Guyot *et al.* (2021) analyzed images from airborne LiDAR data within a 200 km² area in Brittany, a region with a unique megalithic heritage. They used a deep R-CNN approach to partition the digital images into meaningful regions and detect topographic anomalies that may be further analyzed. Bonhage *et al.* (2021) studied pre-industrial age charcoal and its legacy effects on the soil. Automatic site detection is challenging, and improvements have been made by combining LiDAR and multispectral satellite data (Berganzo-Besga *et al.*, 2021). Multispectral cameras extend the spectra of colors from the visible to wavelengths invisible to the human eye, such as infrared and ultraviolet, resulting in richer information in the remote-sensed images (Liu *et al.*, 2023b; Argyrou and Agapiou, 2020).

CNNs are now broadly adopted, with examples ranging from Dolmen detection (Marçal *et al.*, 2024) to broadscale LiDAR detection of Maya sites (Character *et al.*, 2024). The field is maturing with efforts to standardize dealing with large data and defining best practices (Casini *et al.*, 2023; Bellat and Scholten, 2024).

Beyond prospection and site analysis, CNNs are also being explored for surveillance of archaeological sites (Mishra and Lourenço, 2024). Fujita *et al.* (2020) fine-tuned a pre-trained mask R-CNN to detect surface deterioration of stone-made archaeological objects. Hatır *et al.* (2021) collected over 2200 images of Yazılıkaya monuments in the Hattusa archeological site to train a CNN network to distinguish deteriorations of the monument. Again, the difficulty is having enough examples of deterioration to train the network.

2.2.4.2 *Classification and typology*

CNNs excel at image recognition tasks, making them ideal for classifying archaeological artifacts, microfossils, or pollens. These properties make CNN well-suited for automatically classifying archeological artifacts or microfossils. After a training phase, the CNN classifies new samples that can be stored digitally and retrieved at will. The first studies on automatic classification show very encouraging results.

Yalov-Handzel *et al.* (2024) trained a CNN with several thousand images from the Lower Paleolithic to the Late Islamic period and demonstrated a top-5 accuracy above 90%. According to the authors, the classification quality already has the potential to support specialists. A related problem is determining possible sites or regions given an artifact of unknown origin. Such an approach is difficult as similar objects are found on multiple sites. The strategy is to learn from errors (Resler *et al.*, 2021).

In such an application, the output is ideally not a single class but several classes with an associated probability. Given a find from unknown provenance, the network predicts its possible origin.

The field of automatic classification is emerging and requires the development of protocols to integrate ML into the working process (Casini *et al.*, 2023). Without clear (verification) protocols, the scientific community will reject automatic classifications, which would be counterproductive, as the network is, among others, a giant memory at the service of specialists.

2.2.4.3 *Digital epigraphy*

During the eruption of the Vesuvius, a collection of papyri stored in a villa was carbonized and so partially preserved. The papyri were found, rolled, and deposited on shelves. Six papyri were donated to Napoleon Bonaparte, who brought them to France. Before trying to read a papyrus, it must be virtually unrolled using software tools, and the patches with ink must be identified. Physically unrolling a papyrus is very difficult and generally leads to its destruction. The virtual unrolling and deciphering of the En-Gedi scrolls proved the feasibility of the unrolling and deciphering approaches (Seales *et al.*, 2011, 2016). The scrolls contained the book of Leviticus. Though the scrolls are made of animal skins and simpler to unroll and decipher than the Herculaneum papyri, deciphering the En-Gedi scrolls was a breakthrough.

The deciphering of the first Herculaneum papyrus was accelerated by making the data open source in the form of a competition. The Herculaneum papyri digital unfolding required a high-resolution method to separate the signals from the rolled pages, achievable by an X-ray phase-contrast tomography (XPCT). This complex technique uses a coherent X-ray source with a defined phase and wavelength (like most lasers but in the X-ray range!) and exploits the variations in the refractive index to enhance the contrast between papyrus material and ink (Brun *et al.*, 2015; Mocella *et al.*, 2015; Bukreeva *et al.*, 2016; Parker *et al.*, 2019).

The ink detection algorithm combined ML using a ResNet model to detect ink and human reading of the ink patches (Marchant, 2023; Nicolardi *et al.*, 2024). The ML tool identified the presence of ink in a yes/no task using supervised learning.

Interestingly, learning started from a few pieces of virtually unrolled papyri and improved with ink detection. The number of samples increased with learning, creating a positive reinforcement. In this application, the output is either ink or no ink; therefore, the network did not need much training data, and the ResNet approach was very successful.

Character recognition with neural networks has been the subject of many studies, including ancient Japanese, Sanskrit, and Maya glyphs (Avadesh and Goyal, 2018; Can *et al.*, 2018; Chen *et al.*, 2020). Some applications are still experimental, like the automatic reading and identification of hieroglyphs or cuneiform writing. Barucci *et al.* (2021) and Guidi *et al.* (2023) apply R-CNN to classify ancient Egyptian hieroglyphs. They envision the segmentation and classification of hieroglyphs from a transliteration perspective and as a starting point for challenging tasks like reading corrupted or erased signs or even identifying a scribe through its style. Most works are still feasibility studies on a limited database. From that perspective, developing a large standardized database for different writings, like the database for cuneiform writing (Homburg *et al.*, 2022; Chen *et al.*, 2023), will permit going beyond feasibility studies!

Besides character recognition, unsupervised learning techniques like clustering are combined with CNN to analyze inscriptions. Cypro-Minoan inscriptions date to the latter part of the second millennium BCE and are mostly found in Cyprus (Corazza *et al.*, 2022), with a few instances in Ugarit or Greece (Tiryns). The inscriptions are probably related to Linear A (Valério, 2018), another undeciphered language. Corazza *et al.* (2022) used an unsupervised learning method based on DeepCluster (Caron *et al.*, 2018) to analyze a corpus of 600 inscriptions on different physical supports. Deep clustering combines a *k*-means clustering algorithm (see Section 1.3.2.1) to extract pseudo-labels from the images and ResNet to classify the data in a unified architecture (Fig. 2.4).

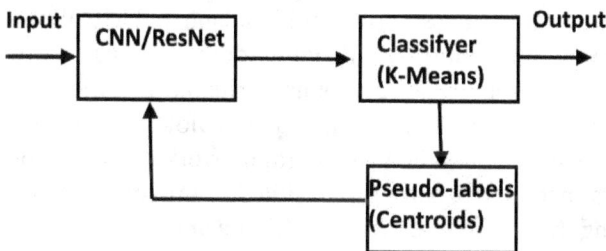

Fig. 2.4. The combination of CNN with a *k*-means classifier generates pseudo-labels. The method is a first step towards a CNN that does not require intensive labeling.

The deep cluster approach represents a significant step forward in unsupervised clustering by combining deep learning and clustering, allowing the processing of images with CNN and clustering the resulting features in an unsupervised manner. The model separated signs on clay tablets from signs attested on other supports. The results show that using different media skews the uniformity of the sign shapes and suggests a unitary, single Cypro-Minoan script rather than the current division into three subgroups. While the model does not attend to deciphering the script, the neural network approach contributes to the discussion on deciphering. The results of this study are very clear, as clustering was found under closer human inspection to correspond to the different supports for the writing. From a broader perspective, DeepCluster slightly reduces the performance gap between supervised and unsupervised learning. There is still much room for improvement. For instance, the k-means clustering furnishes labels without semantic meaning. Now, depending on the problem, clustering without any labels might be difficult to interpret, a problem that may be dealt with using a few truly labeled data to identify possible matches between the labels and the pseudo labels.

2.2.5 *Palynology and paleontology*

Palynology and Paleontology are related application domains of CNN. Romero *et al.* (2020) give a good example of how deep learning can help the work of specialists. The paper is on the automatic analysis of pollen, combining optical super-resolution imaging with deep-learning classification methods. They demonstrate that deep learning increases the speed and accuracy of assessing fossil pollen. The algorithm taxonomically recognizes pollen grains using super-resolution microscope images. The CNN analyzes nanoscale variation in pollen shape, texture, and wall structure for classification. Fossil pollen is abundant, and ML allows experts to focus their time on the most challenging identifications.

Beyond pollen analysis, CNNs are valuable for microfossil studies. The widespread image accessibility and tedious manual identification make microfossils well-suited for neural network classification. A convolution neural network is particularly suited to microfossils, as identifying and counting them is time-consuming. Some applications are very similar to the one encountered in the previous section, like automatic classification (Yu *et al.*, 2023). Liu *et al.* (2023) developed an automatic taxonomic identification for microfossils based on a large Fossil Image Dataset

(>415,000 images) and a CNN to support researchers in identifying the taxa and presenting some alternatives to their analysis. The system obtains results comparable to those of a researcher on some clades (a clade is a group of organisms that consists of a common ancestor and all its descendants). However, the system has difficulties classifying rare specimens.

While microfossils are abundant and well-suited for automated image analysis, applying CNNs to larger and less abundant fossils, such as dinosaur remains, presents unique challenges. Nevertheless, CNNs can still contribute valuable insights to paleontological research, for instance, on dinosaurs. Lallensack *et al.* (2022) discriminate ornithischian and theropod footprints using a CNN architecture on a dataset with more than 1000 footprints. The models performed well in identifying the two types of footprints, but sampling bias remains a major problem for a finer identification. As the example of ink detection in a papyrus showed, focusing on only two classes makes the problem solvable with a CNN without much training data.

2.2.6 *How CNNs connect specialists, companies, and research*

ML is still far from replacing specialized human knowledge in most applications this book covers. Still, it has already reached the point where AI and data analytics permit tackling new questions and supporting specialists with tedious or repetitive tasks. CNNs have impressive capacities for recognizing similar objects that are often rightly classified, even when viewed from different perspectives and distances. On the negative side, a CNN requires many labeled examples in the training phase, which are not always available, causing a large labeling effort. CNNs have difficulties with imbalanced data in the learning dataset, meaning that rare but important instances may have a bad detection or recall rate.

Collaborations between industrial and university labs are important as the most advanced deep learning algorithms often require complex computing infrastructures. The Mask R-CNN technology is an example of how university research took advantage of commercial research and vice versa (Meta open-sources some Mask R-CNN versions). Training very large neural networks from scratch is beyond the scope of researchers interested in specialized applications. They increasingly rely on software made available by some companies providing transparent interfaces.

The number of applications is increasing rapidly, and we have presented some of the best examples showing the potential of CNN.

New neural network architectures regularly extend the range of applications where they perform better than classical approaches. As with any new technique, the results obtained with CNN are scrutinized by the specialists and rightly criticized. A specialist's judging of the network performance is important in understanding the network performance level. Applications of CNNs are generally better accepted than most other architectures by specialists in the field as, in many instances, the results can be relatively easily evaluated by human experts. Letting a neural network or expert system make suggestions is enriching, provided the research community is in the loop. Research should not be left alone to machine or ML specialists, as AI and CNN tend to make mistakes, as men also do, but often different ones. One important role of the expert is to suggest challenging labeled data to check the system classification power.

A difficulty in assessing the quality of the results is that statistics might not convey CNN's true performance well. If the training and validation data do not differ much, then the performance data may be better than the true capabilities of the network on more diverse data. In that context, benchmarking is very important.

Benchmarking is the classical approach to comparing results on a given data set provided to the research community. Benchmarking in the form of competitions has greatly fostered progress by furnishing simple platforms for rapidly identifying and disseminating performant new architectures and ideas and comparing performances under well-defined conditions.

2.2.7 *Validation and interpretability*

The simplest and most used approach to validation is to keep a subset of images for testing and to measure the performances of the trained networks on that subset. This approach is efficient but raises some interesting questions concerning the choice of the testing subset. Should the subset contain different images from the learning subset or even very different images to check the generalization capabilities of the trained networks? How does one measure the distance between the training and testing subset (see Section 2.6 on Siamese networks and Section 3.5.4 on S-BERT for a possible answer)? CNNs are often regarded as a black box, making validation quite difficult. Understanding the features leading to a classification increases confidence in the results.

Much research is on making the classification interpretable. One central approach to tackling that problem is to use interpretable networks (XAI). One of the main approaches consists of looking at intermediary layers to see which neurons and regions of an image are associated with a given output. While this approach may permit removing serious problems, the approach is not systematically used.

Grad-CAM provides a way to visualize which parts of an image contribute to a decision. Recall that in a CNN, each convolutional layer produces feature maps. The first layers detect simple features (like edges or corners), and deeper layers combine these into more complex representations. Grad-CAM analyzes the activations (the values associated with these features) in the penultimate layer. It then computes the gradient of the output class score with respect to the activations. It allows for identifying which features in the penultimate layer are most important for CNN's final decision. The output of Grad-CAM is a heatmap overlaid on the original image, highlighting the regions the CNN focuses on. The low-resolution heatmap does not identify the exact pixels involved, which limits the method.

Several applications implement the Grad-CAM algorithm (Selvaraju *et al.*, 2017) or some variants on Maya glyphs recognition (Can *et al.*, 2018), fossils taxonomic identification (Liu *et al.*, 2023), or the analysis of 3D models of historical buildings (Matrone *et al.*, 2023). The reader is encouraged to see illustrations of Grad-CAM in the references.

There are alternative approaches to Grad-CAM. In some applications, one searches to determine which pixel contributes most to the decision on a class. The prediction is propagated backward to estimate the contribution of each pixel. This approach was followed by El-Hajj *et al.* (2023). They analyzed a database of images extracted from a corpus of 350 early textbooks based on the Tractatus de Sphaera by Johannes de Sacrobosco (circa 1256), containing many illustrations of mathematical instruments. Mathematical instruments in the CNN-based XAI were found to be characterized by the presence of a scale or a graduation. This observation explained the network's failure to correctly classify mathematical instruments without a scale. With that example, one understands that interpretability significantly enhances the possibilities for validating the results of a network.

In another approach, the activation values for different images can be used to compute the distance between them. Dimension reduction algorithms like MDS or t-SNE (see Section 1.3.1) reduce the data to a 2D representation on which one may determine how well classes separate and discover outliers or images not fitting well to a given class. For instance, Karaderi *et al.* (2022) trained a ResNet to recognize foraminifers and applied the above strategy for interpretation. The authors state that such an approach greatly facilitates the verification of the results and makes the interaction with the specialists more intuitive.

2.3 U-Nets and Their Applications

In the CNN architecture in Fig. 2.1, the network output is typically a class ("cat", "dog"). In many applications, one would like the output to be the same size as the input image, for instance, in an image segmentation task to identify a tumor. A specialized U-net architecture extends CNN to such tasks, with important medical applications to help doctors analyze damaged tissue images. The network got its name from its U-shape form, as seen in Fig. 2.5.

Ronneberger *et al.* (2015) showed that some complex image classification tasks do not need many labeled training images. They train the network with multiple views of the same image but at a different scale, translating and rotating some parts. This technique is called data augmentation. The output of the U-net is typically a transformed image of comparable size to the input image.

One part of the U-net is the contracting part (the encoder), while the second half is the expanding path or the decoder. U-nets use skip connections that pass the high-resolution features from the original image or the encoder path to the decoder, allowing for better image recovery. In the

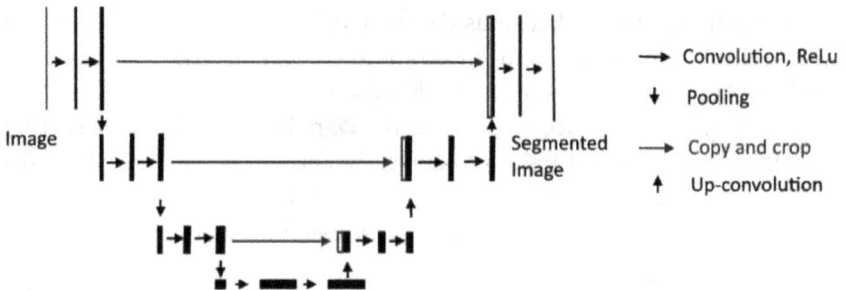

Fig. 2.5.   U-net architecture. Each black box corresponds to a multi-channel feature map, and white boxes represent concatenated data. The arrows denote the different operations.

original design, the number of filters doubles at each instance of downsampling. The signal is then upsampled starting from the lowest layer and concatenated with high-resolution features from the contracting path, helping localize the image features. The network is computationally very efficient and performs very well. In 2016, the U-net won several competitions and challenges on image segmentation by large margins. In real applications, U-Net performs very well. Bundzel *et al.* (2020) compared U-Net and Mask R-CNN for semantic segmentation (identifying areas of ancient construction activities and remnants of ancient Maya buildings). The U-Net-based model performed better. In the work of Fenu *et al.* (2020), the authors exploit a U-Net to analyze mosaics. Each segmented region precisely corresponds to mosaic tesserae. Another approach to the same problem combines a U-net with a standard segmentation algorithm (Felicetti *et al.*, 2021) to fine-tune the result. This research aims to optimize the time-consuming procedure of tesserae segmentation using deep learning and image processing techniques to obtain a digital mosaic replica.

A U-net is quite good at detecting deviations from a standard situation, and some interesting applications focus on identifying defects, for instance, in a production lot of a manufactured product. U-Net learns a compressed representation of the input data and then reconstructs the original input from this representation. If the reconstructed output image is very different from the input image, one may infer that the image has features that were not well captured during learning. The technique was also applied to murals in Han Tombs (Feng *et al.*, 2014) to detect deterioration.

U-nets find applications in paleontology to segment microfossil images (Carvalho *et al.*, 2020). The neural network facilitates the acquisition of statistics on the morphology of the microfossils. In a similar line of research, Beaufort *et al.* (2022) implemented a neural network to process over 7 million specimens. They integrated the results in their analysis pipeline of the time evolution of coccolith morphology. Coccoliths are the calcium-carbonate residue of single-cell plankton. Based on CNN's classification, they show that during the last 2.8 million years, the morphological evolution of fossils has been correlated to climate change, mainly determined by Earth's orbital eccentricity.

This overview shows that ResNet and U-net share many architectural features and applications. ResNet is mostly about classification, and U-net is about segmentation. U-net can also be used for classification through post-processing or adding a layer.

2.4 Generative Networks and Their Applications

Artificial intelligence can create new images almost indistinguishable from real ones using generative adversarial networks (GANs), a class of ML algorithms designed by Goodfellow *et al.* (2014). To achieve this, GANs employ an adversarial process involving two neural networks, the generator and the discriminator, competing against each other (Fig. 2.6).

Specifically, the discriminator D is trained to distinguish real images from those generated by the generator G, accordingly labeling them as real or fake. Conversely, generator G aims to produce synthetic images that are realistic enough to fool the discriminator. This adversarial process can be formalized as a min–max game (Zhang *et al.*, 2024). The generator tries to maximize a function V(G, D) representing the discriminator's error rate. Conversely, the discriminator tries to minimize it. Both networks update their weights iteratively based on their performance in this game. The generator is optimal when its generated data distribution closely resembles the real distribution, effectively fooling the discriminator in 50% of the cases.

Beyond image generation, GANs have demonstrated remarkable learning capabilities in diverse applications. For instance, a GAN has been shown to "teach itself" to play games without explicit knowledge of the game's rules (Kim *et al.*, 2020). By training a generator to produce moves

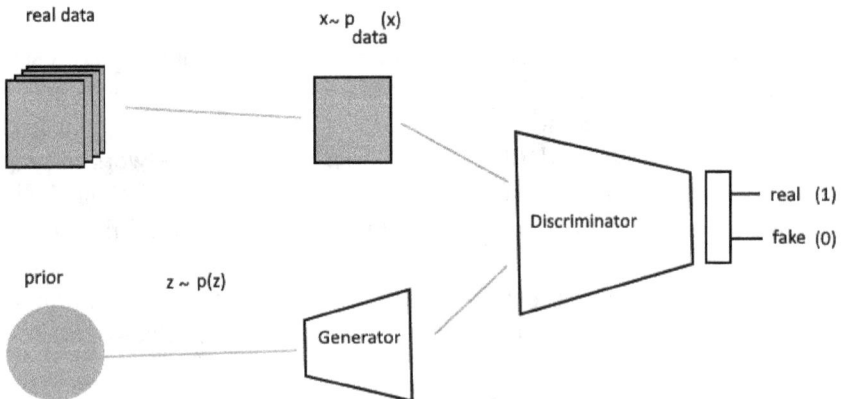

Fig. 2.6. Schematic principle of a GAN. A generator transforms a vector drawn from a random distribution into an image or sound. The discriminator is trained to distinguish between fake and true images.

that can fool a discriminator trained on real games, the generator can essentially learn to play the game without needing to be explicitly taught the rules.

Suppose the data for learning are quite specialized, for instance, containing only human faces. In that case, the faces generated by the best systems are so realistic that they often fool both the discriminator and humans. As initially developed, GAN had some flaws, such as people with two eyes of different colors. Such problems are now mostly solved. CycleGAN is an extension of GAN. In CycleGAN (Zhu *et al.*, 2017), a generator, G, learns the mapping from an image X into its transform Y. A second generator, F, learns the inverse transform. A consistency loss term tells if the inverse image transformation (Y→X) is consistent with the original images.

The possibility of creating deepfakes is a concern for most AI researchers, especially regarding mass manipulation and privacy, but some applications are potentially useful. Let us give a few examples of some well-known applications. Landscape paintings of famous artists are supplied to the system, as landscapes from Van Gogh. After learning the system, landscape pictures were transformed into paintings in van Gogh's style (Liu *et al.*, 2018). After being supplied with noise, the decoder generates a new image in the Van Gogh style. The system generates a different picture when another instance of random noise is provided. The idea is further developed in latent diffusion networks (Section 2.5.1). One speaks of style transfer, an application explored by Gatys *et al.* (2016). This research direction furnishes very spectacular results not only in image but also in speech processing.

Inpainting is one of the main techniques for digital object restoration. U-nets and GAN are broadly used, but latent diffusion models (LDMs) (Section 2.4) are taking the lead. Completing an image while keeping the texture, like the undamaged parts of a ceramic, a painting, or a piece of tissue, is even more difficult than style transfer. Let us discuss GAN in that context. In a digital restoration, the image alterations should, at best, apply to the region to be restored and not to the rest of the image. Some systems require delimiting the image section that should be digitally repaired. In advanced models, learning is triggered to guide the model in generating lost figure contours within the damaged region (Deng *et al.*, 2023).

Some networks specialize in eliminating cracks in murals or other characteristic problems (Yu *et al.*, 2021; Li *et al.*, 2021; Wu *et al.*, 2023). Very encouraging results were obtained on the digital restoration of

Fig. 2.7. Digital restoration of a mural from the temples of Wutaishan, a UNESCO Heritage Site, with a GAN (Cao *et al.*, 2020). Creative Commons license CC-BY. The sword is digitally restored.

murals using GAN-based models (Deng *et al.*, 2023; Wu *et al.*, 2023) on real damages. Figure 2.7 shows an example from the temples of Wutaishan, a UNESCO Heritage Site in the Shanxi Province in China. The task would have been simple for a human operator.

Digital restoration with neural networks still has much room for improvement. Research is necessary to understand theoretically better how to teach a system to recognize defects and make meaningful suggestions for restoration (Bau *et al.*, 2019; Xia *et al.*, 2022). Much work focuses on the Dunhuang murals in Gansu province (China), an oasis on the Silk Road. A comprehensive dataset for digital restoration is provided to researchers to test their method for image restoration (color, impainting) or image enhancement techniques (super-resolution) that may reveal details that are not easily visible due to damage (Xu *et al.*, 2024). Digital restoration methods are not limited to 2D objects and neural networks. The reader will find a good overview of the different techniques in Basu *et al.* (2023), as the topic is too vast to be covered in that section.

Ferreira-Chacua and Koeshidayatullah (2023) report on an interesting approach based on a GAN and a visual transformer (ViT), a neural network architecture for vision inspired by transformers (see Chapter 3) in natural language processing (NLP) (Dosovitskiy, 2020). The GAN generates realistic, high-resolution synthetic images of microfossils based on low-resolution and sometimes incomplete images. These images "augment" the labeled microfossil image database necessary for learning with neural networks. Image generation can enhance the database required for learning and improve an image's quality through denoising and enhancing an image resolution. While these techniques are incredibly exciting, one must remember that their purpose is primarily to create new images; therefore, their use in science requires good validation protocols.

One observes a shift from GAN to other approaches. The transformer technology described in Chapter 3 or the diffusion models described in the next section are often used in generative networks, like SORA, Imagen (Saharia *et al.*, 2022), or DALL-E (Yang *et al.*, 2024a, 2024b). Transformers have the great strength of being multimodal. They process different input types (text, images, and sound).

2.5 Latent Diffusion Models

Diffusion models are a new class of neural networks gaining prominence in generative AI and have some advantages on GAN. They offer a novel approach to synthesizing images by reversing a process of gradual noise injection. The original idea was to apply ideas in statistical physics, a mature field, to generative models (Sohl-Dickstein *et al.*, 2015). These models gradually add noise to an image until it becomes completely random, and then they learn to reverse this process to restore the original image. Each reconstruction step is small and tractable (Ho *et al.*, 2020), making the learning process more stable and less prone to failure than GANs. After learning, the system has the potential to generate an image from noise using the reconstruction stage.

The first models were very slow. A significant advancement in diffusion models came with the introduction of LDMs by Rombach *et al.* (2022). These models operate in a compressed latent space, making them more efficient. As illustrated in Fig. 2.8, a LDM gradually adds noise to an image. It uses a U-net architecture to generate a similar image from noise by removing noise step-by-step. Many advanced systems incorporate the CLIP model to enable text-guided image generation. CLIP associates text and images using a dual encoder, the one processing text and the other the images, learning the relationships between text and images. When a user provides a text prompt, CLIP (Radford *et al.*, 2021) transforms it into an embedding vector. The CLIP embedding "guides" or

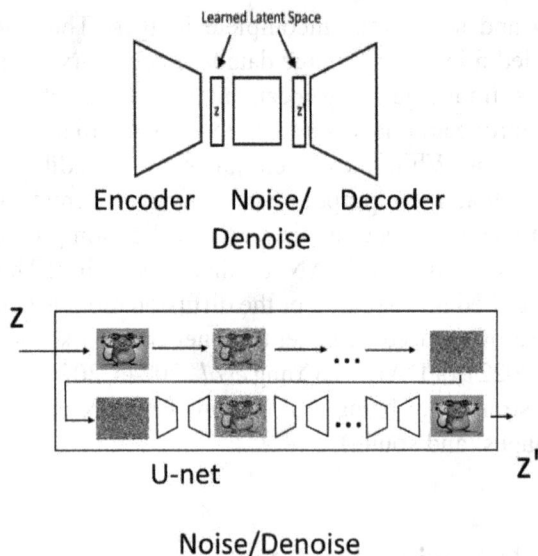

Fig. 2.8. LDM architecture cascading U-net (top) to reconstruct the original image progressively blurred by white noise. The system learns to reconstruct images starting from noise using one U-net at each reconstruction stage.

conditions the model during training by providing the embedding as an additional input to the U-net. At each step of the denoising process, the CLIP embedding of the intermediate output is compared to the desired embedding. This comparison guides the U-Net's denoising process, ensuring the generated image aligns with the text.

Diffusion models have applications, including restoring damaged artwork and analyzing ancient inscriptions. Shao *et al.* (2023) use a modified diffusion model to digitally restore murals and parietal art with real damages. The results are quite convincing. Much progress is still required to make digital restoration with neural networks the standard method.

Some applications combine a CNN with a GAN. Oracle bone inscriptions on turtle shells and animal bones are an important source of information on the Shang Dynasty. Slightly above one-third of the 4500 registered characters have been translated yet. Contrary to English languages, which can be processed using NLP and methods like BERT (Chapter 3), the characters in the inscription must be treated as images. AI methods recognize characters already deciphered on shells and bones (Meng *et al.*, 2018) using a CNN and a GAN to generate new instances of a character to augment the database used for learning (Yue *et al.*, 2022). A new approach (Guan *et al.*, 2024) based on diffusion models tries deciphering unknown characters. The idea is to use old characters in oracle bones and a tuned diffusion model to reconstruct a modern character. Despite the long character evolution over time, the authors claim that some specific local features are preserved and help the system generate proposals for unknown characters after a training phase on translated characters. While the direction is interesting, the future will tell if such studies will contribute significantly to deciphering oracle bones.

2.6 Siamese Networks

The reliance of CNN on extensive labeled datasets is a serious issue. CNNs require too many training examples to be implemented in many applications. Recent advancements in neural network architectures are addressing this challenge through innovative approaches. New neural network architectures tackle these issues in systems capable of generating many labeled images from a few pictures or in architectures requiring fewer examples to learn. Labeling data is labor-intensive, and the demand for networks that can effectively learn from limited data is high. With their ability to learn from little data, Siamese networks present a potential solution to this challenge.

Figure 2.9 shows their structure. Two identical networks share their design, parameters, and weights. In a Siamese network, the encoders have the same weights ($W1 = W2$). The main idea is to make the system learn if two images are from the same class, which is simpler than classification. Labeled versions of an image are generated by deforming them or adding sub-images (Leal-Taixé *et al.*, 2016). Siamese networks can learn, with that approach, meaningful representations without many images (Chen and He, 2021). Siamese Networks are well suited to comparison tasks but are often less performant than CNN in classification.

Two studies use Siamese networks to study mimicry in nature and try dating the appearance of mimicry during evolution (Xu *et al.*, 2022; Fan *et al.*, 2022). As an example of mimicry, debris-carrying insects use various materials to camouflage, protect, or deceive. The study suggests that mimicry is already widespread in the mid-Cretaceous, including plant mimesis in a cricket and debris-carrying camouflage in a toad bug and bark lice. In those examples, the labeled database would not have been large enough for a CNN.

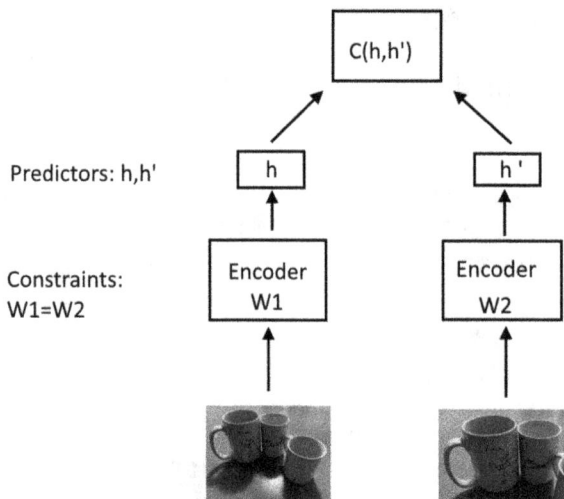

Fig. 2.9.   Sketch showing a Siamese neural network. The two encoders are identical and use the same weights, $W = W1 = W2$. The output C is a measure of similarity between the two images.

Siamese networks are sometimes referred to as a contrastive approach, in which the network decides if two images belong to the same class. In some versions, the Siamese network can solve the task of deciding which of two images is closest to a reference (anchor) image.

Pirrone *et al.* (2021) use Siamese networks to identify which Papyri fragments belong together, a tedious task for a human. The authors compare the problem to "having a bag of puzzles in which all the pieces are mixed and without knowing if all the pieces are present". Training is done using artificially generated fragments of known papyri (e.g., digitally cutting up scanned papyri) and, therefore, does not need any labeling. The network learns to distinguish fragments that belong to the same papyri. They use a contrastive loss function during learning. The contrastive loss function corresponds to the network output error on a proximity measure between two segments. A quite encouraging result was achieved on real fragments, with a 70% top-1 success rate.

Finally, we mention a nice feature of Siamese networks, with applications in image processing and NLP. Siamese networks provide an objective and quantifiable measure of similarity. After learning, the distance between images or text can be computed from their representation in a low-dimension space.

2.7 Complement: The Components of a CNN

The discussion refers to Fig. 2.1 in Section 2.2, which we have copied here to help the reader.

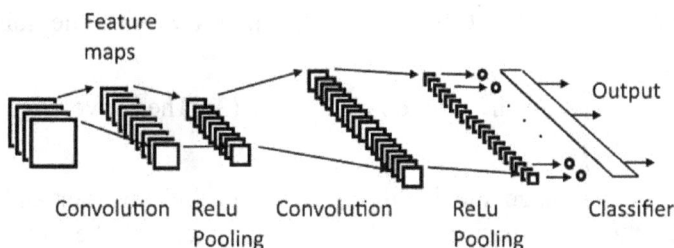

This section explains the function of the main components of a CNN with simple examples.

2.7.1 *Convolutional filter*

A convolution filter is a small matrix that extracts image features. Convolution filters may detect specific patterns within an image, such as edges, textures, or more complex shapes. For example, consider a small image coded by the matrix M.

$$M = \begin{pmatrix} 1 & 2 & 1 \\ 1 & 1 & 3 \\ 1 & 4 & 3 \end{pmatrix}$$

and the convolution filter $F = \begin{pmatrix} 1 & -1 \\ -1 & 1 \end{pmatrix}$. Convolution differs from matrix

multiplication; in a convolution filter, the multiplication is done term by term. The matrix M' after convolution is

$$M' = \begin{pmatrix} -1 & 3 \\ 3 & -3 \end{pmatrix}.$$

Let us consider the first term resulting from the convolution of the subma-

trix $S_{11} = \begin{pmatrix} 1 & 2 & . \\ 1 & 1 & . \\ . & . & . \end{pmatrix}$ with F. The output equals $1 - 2 + 3 - 3 = -1$, the

upper left term of M'. Convoluting the matrix F with the submatrix

$S_{12} = \begin{pmatrix} . & 2 & 1 \\ . & 1 & 3 \\ . & . & . \end{pmatrix}$ furnishes the top right value (3). The convolution opera-

tor slides on the image, making the convolution shift-invariant and, therefore, robust against a translation of the image. In a CNN, the values of the
convolution filters are learned during the training phase. These filters play
a crucial role with their versatility and ability to detect specific patterns
within an image, such as edges, textures, or more complex shapes. Some
filters may be edge detectors. Understanding how an edge filter works is
quite instructive in getting an intuition on how a CNN extracts features.
Here is an example:

$$M = \begin{pmatrix} 0 & -1 & 0 \\ -1 & 4 & -1 \\ 0 & -1 & 0 \end{pmatrix}$$

The output is zero if the filter is applied to a region with a constant
value. For an edge, one obtains non-zero values. In CNN, the values of the
filters belong to the learnable parameters determined during learning to
maximize the success of classifying the labeled images. Let us note that
care must be taken to the boundary of the images by adding rows and
columns of pixels at the boundary (padding).

2.7.2 Pooling

In the context of neural networks, pooling means that one applies a filter
keeping the largest value or a low-pass filter to the output of the convolution filters, and the resulting output is sub-sampled. A low-pass filter aver-
ages or blurs an image. The matrix $LP = \begin{pmatrix} 1/4 & 1/4 \\ 1/4 & 1/4 \end{pmatrix}$ is a simple example
of a low-pass filter. The filtering process is identical to the above example,

which uses a convolution filter. The matrix *LP* applied to S_{11} furnishes the value 5/4. The role of the low-pass filter is to average neighboring values before downsampling. Convolutional layers can have a stride greater than 1. In that case, the filter jumps multiple pixels (the stride gives the jump distance), effectively reducing the output dimensions.

2.7.3 *Nonlinear operator*

A rectified linear unit (ReLu) is a nonlinear transformation defined as

$$f(x) = \max(x,0).$$

The ReLu operator is a simple and efficient nonlinear operator that transfers the input value unchanged to the output and sets all negative values to zero (Fig. 2.10). Nonlinearity follows from setting negative values to zero. The cut limit (here zero) is generally a learned parameter. Other activation functions like Sigmoid and Tanh are implemented instead of ReLu.

2.7.4 *Classifier*

A simple classical classifier, for instance, a fully connected one-layer network, furnishes the result after several stages: convolution stage, pooling, and nonlinear processing.

2.7.5 *Learning*

Learning corresponds to determining the best weights that describe the labeled samples. Backpropagation is the classical approach. The

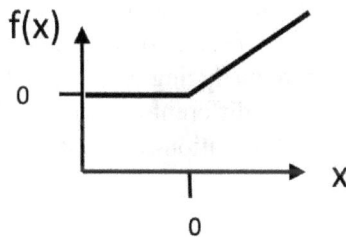

Fig. 2.10. Nonlinear ReLu activation function. The derivative is one above the threshold and zero below. At the threshold, one attributes the derivative to zero for practical purposes.

classification error or loss L corresponds to the difference between the output vector and the desired class (a vector with all zeros except one element equal to one). Backpropagation determines how much a small change in the filter weights and the bias affects the classification error (loss: L). Given the weights (or elements of the matrix) of a convolution kernel and its output, one uses the chain rule for derivation to compute the gradients from the bottom layer iteratively. If a loss function equals the square difference between the expected and the obtained values, one has $\left(\frac{\partial L}{\partial y} = 2(y - y_{exp}) \right)$. The gradient of the ReLu is either one or zero (Fig. 2.10). The gradients serve to update the weights and biases using an optimization algorithm like gradient descent, which aims to minimize the loss. The network is trained or fine-tuned with standard tools (TensorFlow, PyTorch, or Keras).

The success of the convolutional network architecture has baffled the specialists even more because their success was tough to figure out mathematically. The work by Stephane Mallet shed some light on why convolutional networks are so efficient in learning. Several features of deep convolution networks can be related to multiresolution analysis and wavelet theory. A wavelet filter is used in signal and image processing to analyze the data at different scales or resolutions. Mallat (2016) explains some of the properties of convolution networks using a wavelet filter as a convolution filter. He shows the necessity of combining a convolution, a nonlinear, and a low-pass filter stage (so-called pooling). Averaging the filtered data with a low-pass filter makes the filter robust against small image deformations. The low-pass filter 'washed out' the data, resulting in some apparent information loss. A nonlinear filter introduces frequencies that correspond to frequency differences in the unfiltered signal (for instance, a rectifier $\rho(x) = \max(x + b, 0)$. In other words, nonlinear filtering permits the recovery of some frequencies that the low-pass filter would filter out without it. By cascading several modules, the signal is analyzed at lower frequencies at each stage until no signal energy remains. The many layers are equivalent to analyzing the image at different resolutions, capturing the main features at different resolutions. In other words, combining modules results in a translation-invariant multiresolution analysis.

References

Argyrou, A., & Agapiou, A. (2022). A review of artificial intelligence and remote sensing for archaeological research, *Remote Sensing*, 14(23), p. 6000.

Avadesh, M., & Goyal, N. (2018). Optical character recognition for Sanskrit using convolution neural networks, *Proceedings of the 13th IAPR International Workshop on Document Analysis Systems (DAS)*, (IEEE Computer Society) pp. 447–452.

Barucci, A., Canfailla, C., Cucci, C., Forasassi, M., Franci, M., Guarducci, G., ... & Argenti, F. (2022). Ancient Egyptian hieroglyphs segmentation and classification with convolutional neural networks, *International Conference Florence Heri-Tech: The Future of Heritage Science and Technologies*, (Springer International Publishing) pp. 126–139.

Basu, A., Paul, S., Ghosh, S., Das, S., Chanda, B., Bhagvati, C., & Snasel, V. (2023). Digital restoration of cultural heritage with data-driven computing: A survey, *IEEE Access*, 11, pp. 53939–53977.

Bau, D., Zhu, J. Y., Wulff, J., Peebles, W., Strobelt, H., Zhou, B., & Torralba, A. (2019). Seeing what a gan cannot generate, *Proceedings of the IEEE/CVF International Conference on Computer Vision*, pp. 4502–4511.

Beaufort, L., Bolton, C. T., Sarr, A. C., Suchéras-Marx, B., Rosenthal, Y., Donnadieu, Y., ... & Tetard, M. (2022). Cyclic evolution of phytoplankton forced by changes in tropical seasonality, *Nature*, 601(7891), pp. 79–84.

Bellat, M., & Scholten, T. (2024). Automated features detection in archaeology: Standardisation in the area of big data, *CAA51st Across the Horizon*.

Berganzo-Besga, I., Orengo, H. A., Lumbreras, F., Carrero-Pazos, M., Fonte, J., & Vilas-Estévez, B. (2021). Hybrid MSRM-based deep learning and multitemporal sentinel 2-based machine learning algorithm detects nearly 10k archaeological tumuli in North-Western Iberia, *Remote Sensing*, 13(20), p. 4181.

Bickler, S. H. (2021). Machine learning arrives in archaeology, *Advances in Archaeological Practice*, 9(2), pp. 186–191.

Bonhage, A., Eltaher, M., Raab, T., Breuß, M., Raab, A. and Schneider, A. (2021). A modified mask region-based convolutional neural network approach for the automated detection of archaeological sites on high-resolution light detection and ranging-derived digital elevation models in the North German Lowland, *Archaeological Prospection*, 28(2), pp. 177–186.

Brun, E., Cotte, M., Wright, J., Ruat, M., Tack, P., Vincze, L., ... & Mocella, V. (2016). Revealing metallic ink in Herculaneum papyri, *Proceedings of the National Academy of Sciences*, 113(14), pp. 3751–3754.

Bukreeva, I., Mittone, A., Bravin, A., Festa, G., Alessandrelli, M., Coan, P., ... & Cedola, A. (2016). Virtual unrolling and deciphering of Herculaneum papyri by X-ray phase-contrast tomography, *Scientific Reports*, 6(1), p. 27227.

Bundzel, M., Jaščur, M., Kováč, M., Lieskovský, T., Sinčák, P., & Tkáčik, T. (2020). Semantic segmentation of airborne lidar data in Maya archaeology, *Remote Sensing*, 12(22), p. 3685.

Can, G., Odobez, J. M., Gatica-Perez, D. (2016). Evaluating shape representations for Maya glyph classification, *Journal of Computing Culture Heritage*, 9, pp. 1–26.

Cao, J., Zhang, Z., Zhao, A., Cui, H., & Zhang, Q. (2020). Ancient mural restoration based on a modified generative adversarial network, *Heritage Science*, 8, pp. 1–14.

Caron, M., Bojanowski, P., Joulin, A., & Douze, M. (2018). Deep clustering for unsupervised learning of visual features, *Proceedings of the European Conference on Computer Vision (ECCV)*, pp. 132–149.

Carvalho, L. E., Fauth, G., Fauth, S. B., Krahl, G., Moreira, A. C., Fernandes, C. P., & Von Wangenheim, A. (2020). Automated microfossil identification and segmentation using a deep learning approach, *Marine Micropaleontology*, 158, p. 101890.

Casini, L., Marchetti, N., Montanucci, A., Orrù, V., & Roccetti, M. (2023). A human-AI collaboration workflow for archaeological sites detection, *Scientific Reports*, 13(1), p. 8699.

Caspari, G., & Crespo, P. (2019). Convolutional neural networks for archaeological site detection–Finding "princely" tombs, *Journal of Archaeological Science*, 110, p. 104998.

Character, L., Beach, T., Inomata, T., Garrison, T. G., Luzzadder-Beach, S., Baldwin, J. D., ... & Ranchos, J. L. (2024). Broadscale deep learning model for archaeological feature detection across the Maya area, *Journal of Archaeological Science*, 169, p. 106022.

Chen, D., Agarwal, A., Berg-Kirkpatrick, T., & Myerston, J. (2023). CuneiML: A cuneiform dataset for machine learning, *Journal of Open Humanities Data*, 9(1), pp. 30–41.

Chen, L., Lyu, B., Tomiyama, H., & Meng, L. A (2020). Method of Japanese ancient text recognition by deep learning, *Proceedings of the Computer Science*, 174, pp. 276–279.

Corazza, M., Tamburini, F., Valério, M., & Ferrara, S. (2022). Unsupervised deep learning supports reclassification of Bronze Age Cypriot writing system, *PLoS One*, 17(7), p. e0269544.

Deng, X., & Yu, Y. (2023). Ancient mural inpainting via structure information guided two-branch model, *Heritage Science*, 11(1), p. 131.

Dosovitskiy, A., Beyer, L., Kolesnikov, A., Weissenborn, D., Zhai, X., Unterthiner, T., ... & Houlsby, N. (2020). An image is worth 16x16 words: Transformers for image recognition at scale, arXiv preprint arXiv:2010.11929.

El-Hajj, H., Eberle, O., Merklein, A., Siebold, A., Shlomi, N., Büttner, J., ... & Valleriani, M. (2023). Explainability and transparency in the realm of digital humanities: toward a historian XAI, *International Journal of Digital Humanities*, 5(2), pp. 299–331.

Fan, L., Xu, C., Jarzembowski, E. A., & Cui, X. (2022). Quantifying plant mimesis in fossil insects using deep learning, *Historical Biology*, 34(5), pp. 907–916.

Felicetti, A., Paolanti, M., Zingaretti, P., Pierdicca, R., & Malinverni, E. S. (2021). Mo. Se.: Mosaic image segmentation based on deep cascading learning, *Virtual Archaeology Review*, 12(24), pp. 25–38.

Feng, J., Zhao, F., & Li, S. (2014). On-site conservation of the tomb mural of the Western Han dynasty at Xi'an University of Technology, Xi'an, China, *Studies in Conservation*, 59 (sup1), pp. S40–S43.

Fenu, G., Medvet, E., Panfilo, D., & Pellegrino, F. A. (2020). Mosaic images segmentation using U-net, *Proceedings of the 9th International Conference on Pattern Recognition Applications and Methods* (Scitepress) pp. 485–492.

Ferreira-Chacua, I., & Koeshidayatullah, A. (2023). ForamViT-GAN: Exploring new paradigms in deep learning for micropaleontological image analysis, *IEEE Access*.

Fujita, H., Itagaki, M., Ichikawa, K., Hooi, Y. K., Kawano, K., & Yamamoto, R. (2020). Fine-tuned pre-trained mask R-CNN models for surface object detection, arXiv preprint arXiv:2010.11464.

Gatys, L. A., Ecker, A. S., & Bethge, M. (2016). Image style transfer using convolutional neural networks, *Proceedings of the IEEE Conference on Computer Vision and Pattern Recognition*, pp. 2414–2423.

Girshick, R., Donahue, J., Darrell, T., & Malik, J. (2014). Rich feature hierarchies for accurate object detection and semantic segmentation, *Proceedings of the IEEE Conference on Computer Vision and Pattern Recognition*, pp. 580–587.

Goodfellow, I., Pouget-Abadie, J., Mirza, M., Xu, B., Warde-Farley, D., Ozair, S., ... & Bengio, Y. (2014). Generative adversarial nets, *Advances in Neural Information Processing Systems*, p. 27.

Guan, H., Yang, H., Wang, X., Han, S., Liu, Y., Jin, L., ... & Liu, Y. (2024). Deciphering oracle bone language with diffusion models, arXiv preprint arXiv:2406.00684.

Guidi, T., Python, L., Forasassi, M., Cucci, C., Franci, M., Argenti, F., & Barucci, A. (2023). Egyptian hieroglyphs segmentation with convolutional neural networks, *Algorithms*, 16(2), p. 79.

Guyot, A., Lennon, M., & Hubert-Moy, L. (2021). Objective comparison of relief visualization techniques with deep CNN for archaeology, *Journal of Archaeological Science: Reports*, 38, p. 103027.

Hatır, E., Korkanç, M., Schachner, A., & Ince, I. (2021). The deep learning method applied to the detection and mapping of stone deterioration in open-air sanctuaries of the Hittite period in Anatolia, *Journal of Cultural Heritage*, 51, pp. 37–49.

He, K., Zhang, X., Ren, S., & Sun, J. (2016), Deep residual learning for image recognition, *Proceedings of the IEEE Conference on Computer Vision and Pattern Recognition*, pp. 770–778.

Ho, J., Jain, A., & Abbeel, P. (2020). Denoising diffusion probabilistic models, *Advances in Neural Information Processing Systems*, 33, pp. 6840–6851.

Homburg, T., Zwick, R., Mara, H., & Bruhn, K. C. (2022). Annotated 3D-models of cuneiform tablets, *Journal of Open Archaeology Data*, 10, pp. 1–8.

Jamil, A. H., Yakub, F., Azizan, A., Roslan, S. A., Zaki, S. A., & Ahmad, S. A. (2022). A review on Deep Learning application for detection of archaeological structures, *Journal of Advanced Research in Applied Sciences and Engineering Technology*, 26(1), pp. 7–14.

Karaderi, T., Burghardt, T., Hsiang, A. Y., Ramaer, J., & Schmidt, D. N. (2022). Visual microfossil identification via deep metric learning, *International Conference on Pattern Recognition and Artificial Intelligence* (Springer International Publishing), pp. 34–46.

Kattenborn, T., Leitloff, J., Schiefer, F., & Hinz, S. (2021). Review on convolutional neural networks (CNN) in vegetation remote sensing, *ISPRS Journal of Photogrammetry and Remote Sensing*, 173, pp. 24–29.

Kaul, V., Enslin, S., & Gross, S. A. (2020). History of artificial intelligence in medicine, *Gastrointestinal Endoscopy*, 92(4), pp. 807–812.

Kim, S. W., Zhou, Y., Philion, J., Torralba, A., & Fidler, S. (2020). Learning to simulate dynamic environments with gamegan, *Proceedings of the IEEE/CVF Conference on Computer Vision and Pattern Recognition*, pp. 1231–1240.

Lallensack, J. N., Romilio, A., & Falkingham, P. L. (2022). A machine learning approach for the discrimination of theropod and ornithischian dinosaur tracks, *Journal of Royal Society Interface*, 19(196), p. 20220588.

Lawal, O. M. (2023). YOLOv5-LiNet: A lightweight network for fruits instance segmentation, *PLoS One*, 18(3), p. e0282297.

LeCun, Y., Bengio, Y., Hinton, G. (2015). Deep learning, *Nature*, 521(7553), pp. 436–444.

Leal-Taixé, L., Canton-Ferrer, C., & Schindler, K. (2016). Learning by tracking: Siamese CNN for robust target association, *Proceedings of the IEEE Conference on Computer Vision and Pattern Recognition Workshops*, pp. 33–40.

Li, J., Wang, H., Deng, Z., Pan, M., & Chen, H. (2021). Restoration of non-structural damaged murals in Shenzhen Bao'an based on a generator–discriminator network, *Heritage Science*, 9, pp. 1–14.

Li, S., Jiao, J., Han, Y., & Weissman, T. (2016). Demystifying resnet, arXiv preprint arXiv:1611.01186.

Liu, H., Michelini, P. N., & Zhu, D. (2018). Artsy-GAN: A style transfer system with improved quality, diversity and performance, *2018 24th International Conference on Pattern Recognition (ICPR)*, pp. 79–84.

Liu, L., Miteva, T., Delnevo, G., Mirri, S., Walter, P., de Viguerie, L., & Pouyet, E. (2023b). Neural networks for hyperspectral imaging of historical paintings: A practical review, *Sensors*, 23(5), p. 2419.

Liu, X., Jiang, S., Wu, R., Shu, W., Hou, J., Sun, Y., ... & Song, H. (2023). Automatic taxonomic identification based on the Fossil Image Dataset (>415,000 images) and deep convolutional neural networks, *Paleobiology*, 49(1), pp. 1–22.

Long, J., Shelhamer, E., Darrell, T. (2015). Fully convolutional networks for semantic segmentation, *Proceedings of the IEEE Conference on Computer Vision and Pattern Recognition*, pp. 3431–3440.

Mallat, S. (2016). Understanding deep convolutional networks, *Philosophical Transactions of the Royal Society A: Mathematical, Physical and Engineering Sciences*, 374(2065), p. 20150203.

Marçal, D., Câmara, A., Oliveira, J., & de Almeida, A. (2024). Evaluating R-CNN and YOLO V8 for megalithic monument detection in satellite images, *International Conference on Computational Science* (Springer Nature Switzerland) pp. 162–170.

Marchant, J. (2023). AI reads text from ancient Herculaneum scroll for the first time, *Nature*. doi: 10.1038/d41586-023-03212-1.

Matrone, F., Felicetti, A., Paolanti, M., & Pierdicca, R. (2023). Explaining AI: understanding deep learning models for heritage point clouds, *ISPRS Annals of the Photogrammetry, Remote Sensing and Spatial Information Sciences*, pp. 207–214.

Mehendale, N. (2020). Facial emotion recognition using convolutional neural networks (FERC), *SN Applied Sciences*, 2(3), p. 446.

Meng, L., Kamitoku, N., & Yamazaki, K. (2018). Recognition of oracle bone inscriptions using deep learning based on data augmentation, *Proceedings of the 2018 Metrology for Archaeology and Cultural Heritage* (MetroArchaeo) pp. 33–38.

Mishra, M., & Lourenço, P. B. (2024). Artificial intelligence-assisted visual inspection for cultural heritage: State-of-the-art review, *Journal Cultural Heritage*, 66, pp. 536–550.

Mocella, V., Brun E., Ferrero C., & Delattre D. (2015). Revealing letters in rolled Herculaneum papyri by X-ray phase-contrast imaging, *Nature Communications*, 6(1), p. 5895.

Navarro, P., Cintas, C., Lucena, M., Fuertes, J. M., Delrieux, C., & Molinos, M. (2021). Learning feature representation of Iberian ceramics with automatic classification models, *Journal of Cultural Heritage*, 48, pp. 65–73.

Nicolardi, F., Parsons, S., Delattre, D., Del Mastro, G., Fowler, R. L., Janko, R., ... & Brent Seales, W. (2024). Revealing text from a still-rolled Herculaneum papyrus scroll (PHerc. Paris. 4), *Zeitschrift fuer Papyrologie und Epigraphik*, 229, pp. 1–13.

Oquab, M., Bottou, L., Laptev, I., & Sivic, J., (2014). Learning and transferring mid-level image representations using convolutional neural networks, *Proceedings of the IEEE Conference on Computer Vision and Pattern Recognition*, pp. 1717–1724.

Parker, C. S., Parsons, S., Bandy, J., Chapman, C., Coppens, F., & Seales, W. B. (2019). From invisibility to readability: Recovering the ink of Herculaneum, *PLoS One*, 14(5), p. e0215775.

Pirrone, A., Beurton-Aimar, M., & Journet, N. (2021). Self-supervised deep metric learning for ancient papyrus fragments retrieval, *International Journal on Document Analysis and Recognition (IJDAR)*, 24(3), pp. 219–234.

Radford, A., Kim, J. W., Hallacy, C., Ramesh, A., Goh, G., Agarwal, S., ... & Sutskever, I. (2021). Learning transferable visual models from natural language supervision, *International Conference on Machine Learning (PLMR)*, pp. 8748–8763.

Resler, A., Yeshurun, R., Natalio, F., & Giryes, R. (2021). A deep-learning model for predictive archaeology and archaeological community detection, *Humanities and Social Sciences Communications*, 8(1), pp. 1–10.

Rombach, R., Blattmann, A., Lorenz, D., Esser, P., & Ommer, B. (2022). High-resolution image synthesis with latent diffusion models, *Proceedings of the IEEE/CVF Conference on Computer Vision and Pattern Recognition*, pp. 10684–10695.

Romero, I. C., Kong, S., Fowlkes, C. C., Jaramillo, C., Urban, M. A., Oboh-Ikuenobe, F., ... & Punyasena, S. W. (2020). Improving the taxonomy of fossil pollen using convolutional neural networks and super-resolution microscopy, *Proc. National Academy of Sciences*, 117(45), pp. 28496–28505.

Ronneberger, O., Fischer, P., & Brox, T. (2015). U-net: Convolutional networks for biomedical image segmentation, *Medical Image Computing and Computer-assisted Intervention–MICCAI 2015, Munich, Germany, October 5-9, 2015, Proc. Part III*, 18 (Springer International Publishing) pp. 234–241.

Saharia, C., Chan, W., Saxena, S., Li, L., Whang, J., Denton, E. L., ... & Norouzi, M. (2022). Photorealistic text-to-image diffusion models with deep language understanding, *Advances in Neural Information Processing Systems*, 35, pp. 36479–36494.

Sarvamangala, D. R., Kulkarni, R.V. (2022). Convolutional neural networks in medical image understanding: A survey, *Evolutionary Intelligence*, 15(1), pp. 1–22.

Seales, W. B., Griffioen, J., Baumann, R., & Field, M. (2011). Analysis of Herculaneum papyri with X-ray computed tomography, *International Conference on Nondestructive Investigations and Microanalysis for the Diagnostics and Conservation of Cultural and Environmental Heritage* (Vol. 7).

Seales, W. B., Parker, C. S., Segal, M., Tov, E., Shor, P., & Porath, Y. (2016). From damage to discovery via virtual unwrapping: Reading the scroll from En-Gedi, *Science Advances*, 2(9), p. e1601247.

Selvaraju, R. R., Cogswell, M., Das, A., Vedantam, R., Parikh, D., & Batra, D. (2017). Grad-cam: Visual explanations from deep networks via gradient-based localization, *Proceedings of the IEEE International Conference on Computer Vision*, pp. 618–626.

Shao, H., Xu, Q., Wen, P., Gao, P., Yang, Z., & Huang, Q. (2023). Building bridge across the time: Disruption and restoration of murals in the wild, *Proceedings of the IEEE/CVF International Conference on Computer Vision*, pp. 20259–20269.

Sohl-Dickstein, J., Weiss, E., Maheswaranathan, N., & Ganguli, S. (2015). Deep unsupervised learning using nonequilibrium thermodynamics, *International Conference on Machine Learning* (PMLR), pp. 2256–2265.

Trier, Ø. D., Reksten, J. H., Løseth, K. (2021). Automated mapping of cultural heritage in Norway from airborne lidar data using faster R-CNN, *International Journal of Applied Earth Observation and Geoinformation*, 95, p. 102241.

Valério, M. (2018). Cypro-Minoan: an Aegean-derived syllabary on Cyprus (and elsewhere), *Paths into Script Formation in the Ancient Mediterranean: Studi Micenei ed Egeo-Anatolici, Nuova Serie*, Supplemento, 1.

Wang, G. G., Cheng, H., Zhang, Y., & Yu, H. (2023). ENSO analysis and prediction using deep learning: A review, *Neurocomputing*, 520, pp. 216–229.

Waziralilah, N. F., Abu, A., Lim, M. H., Quen, L. K., & Elfakharany, A. (2019). A review on convolutional neural network in bearing fault diagnosis, *MATEC Web of Conferences*, 255 (EDP Sciences), p. 06002.

Wu, M., Chang, X., & Wang, J. (2023). Fragments inpainting for tomb murals using a dual-attention mechanism GAN with improved generators, *Applied Sciences*, 13(6), p. 3972.

Wu, W., Kirillov, A., Massa, F., Lo, W.-Y., & Girshick, R. (2019). Detectron2. *Meta AI*, 10, p. 2.

Xia, W., Zhang, Y., Yang, Y., Xue, J. H., Zhou, B., & Yang, M. H. (2022). Gan inversion: A survey, *IEEE Transactions on Pattern Analysis and Machine Intelligence*, 45(3), pp. 3121–3138.

Xu, C., Wang, B., Fan, L., Jarzembowski, E. A., Fang, Y., Wang, H., ... & Engel, M. S. (2022). Widespread mimicry and camouflage among mid-Cretaceous insects, *Gondwana Research*, 101, pp. 94–102.

Xu, Z., Yang, Y., Fang, Q., Chen, W., Xu, T., Liu, J., & Wang, Z. (2024). A comprehensive dataset for digital restoration of Dunhuang murals, *Scientific Data*, 11(1), p. 955.

Yalov-Handzel, S., Cohen, I., & Aperstein, Y. (2024). Comparative analysis of CNN architectures and loss Functions on age estimation of archaeological artifacts, *Journal of Computer Computations in Archaeology*, 7(1), pp. 185–194.

Yang, D., Hu, L., Tian, Y., Li, Z., Kelly, C., Yang, B., Yang, C., & Zou, Y. (2024a). WorldGPT: A Sora-inspired video AI agent as rich world models from text and image inputs, arXiv preprint arXiv:2403.07944.

Yang, H., Lu, Y., & Jiang, H. (2024b). A review of 3D object detection methods for autonomous driving, *Advances in Signal Processing and Artificial Intelligence*, p. 37.

Yu, C., Qin, F., Li, Y., Qin, Z., & Norell, M. (2022). CT segmentation of dinosaur fossils by deep learning, *Frontiers in Earth Science*, 9, p. 805271.

Yu, K., Li, Y., Yan, J., Xie, R., Zhang, E., Liu, C., & Wang, J. (2021). Intelligent labeling of areas of wall painting with paint loss disease based on multi-scale detail injection U-Net, *Optics for Arts, Architecture, and Archaeology VIII*, 11784, (SPIE) pp. 37–44.

Yue, X., Li, H., Fujikawa, Y., & Meng, L. (2022). Dynamic dataset augmentation for deep learning-based oracle bone inscriptions recognition, *ACM Journal of Computing and Cultural Heritage*, 15(4), pp. 1–20.

Zhang, Z., Hua, Y., Wang, H., & McLoone, S. (2024). Improving the fairness of the min-max game in GANs training. *Proceedings of the IEEE/CVF Winter Conference on Applications of Computer Vision*, pp. 2910–2919.

Zhu, J. Y., Park, T., Isola, P., & Efros, A. A. (2017). Unpaired image-to-image translation using cycle-consistent adversarial networks, *Proceedings of the IEEE International Conference on Computer Vision*, pp. 2223–2232.

Part 2

Natural Language Processing and Large Language Models

Chapter 3

Natural Language Processing with Neural Networks

3.1 Introduction

Winter 2022–2023 will be remembered as the time artificial intelligence (AI) changed everybody's daily routine, with large language models (LLMs) revolutionizing our interaction with computers. LLMs can now process spoken or written languages and images. Within a short time, the best models became multimodal with the possibility to process language, sound, and images as well as to generate code for tasks that can be described iteratively in text form. In this chapter, we present the developments chronologically and focus on the applications in the digital humanities. We start with a technique called Word2Vec that analyzes texts without caring about the order of the words as an introduction to word embedding. We then discuss neural networks that consider the order of the words. A famous paper by a Google Team proposed an innovative approach.

The paper entitled "Attention is All You Need" by Vaswani *et al.* (2017) has been extremely influential, with over 150,000 citations in the scientific literature. The 'attention' mechanism is at the basis of LLMs, for instance, in systems capable of translating texts into another language. In the following sections, we explain the attention mechanism in some detail. The original LLM model processed the input sequence in one language and encoded it in a representation that captured semantic and syntactic information. In the second part of the model, the decoder takes the representation from the encoder and generates the next translation word.

Attention is at the core of the neural network architecture that revolutionizes natural language processing (NLP). For instance, Bidirectional Encoder Representations from Transformers (BERT) and generative pre-trained transformer (GPT) form the core of commercial LLMs. Google developed BERT for NLP tasks. The model processes text bi-directionally in both forward and backward directions. Learning is done by teaching the system to guess masked words in a sentence. This approach provides a simple method to learn in a supervised manner without the problem of labeling. In the first versions, learning was also done by taking pairs of sentences, masking one sentence, and making the system find the missing sentence. Learning corresponds to modifying the neurons' weights and other parameters according to the answer's quality. While private companies first released the code of their networks to sustain progress by having an open-source strategy, publications on the latest developments are scarce, a dangerous development for the scientific community. Using the first versions, some communities in many academic fields learned using these tools and understood their potential. Not having access to the latest versions and not being capable of tuning them to their specific needs is very frustrating as the gap between the knowledge in private companies and academic research broadens, especially in applications.

NLP has applications in ancient languages, which we will discuss in the application sections. This chapter introduces the main building blocks of the techniques mentioned in that introduction, mostly self-consistently; it is intended to be understandable without a deep mathematical back-ground (we assume the reader is familiar with matrix multiplications).

3.2 Word Embedding

This section presents the basis for transforming some text into numbers that can be processed with neural networks. Embedding is an important step towards making text tractable for machine learning (ML). The embedding process starts from a very large corpus, such as Wikipedia, or a large, specialized corpus, where the words in the corpus are tokenized. Consider the sentence, "Kepler is a great astronomer". The sentence is split into five tokens: "Kepler", "is", "a", "famous", and "astronomer". In a large corpus, some tokens repeat a lot like "a" or "is" and are quite unspecific, though syntactically important. The words Kepler and astronomers are words that may often appear in connection.

Word embedding represents words as continuous value vectors. Word embedding considers words and ignores their contexts. The main goal of word embedding is to code related words with similar values. Consider a 3D embedding space in which a three-value vector codes a word. Two words with almost identical values are close to each other's in the embedding. The word "astronomy" may be coded as astronomy: [0.1, 0.4, 0.4], while Kepler may have a close coding [0.1, 0.5, 0.4].

Contrary to this simple example, embedding uses a large vector in practice. The original Word2Vec algorithm already used vectors with three hundred dimensions, meaning that three hundred values represent each word. With hundreds of values, one can easily embed "Kepler" close to many other words, such as "scientist" or the city "Prag". As Prag is not only characterized by Kepler, it is already sufficient to make Kepler closer to Prag than to Paris in the embedding.

Let us discuss how these values are computed for words in a dictionary. Word2Vec was developed by a Google team (Mikolov *et al.*, 2013) to transform words into a series of numbers, in other words, as a vector. The project aimed to embed millions of words using a dataset with billions of sentences. Word2Vec learns representations of words based on their co-occurrence patterns in a large corpus of text. Two main architectures are used in Word2Vec: the Continuous Bag of Words (CBOW) and Skip-Gram. Both architectures use a neural network with a single hidden layer to learn meaningful word embeddings.

Figure 3.1 illustrates both algorithms with a simple example. CBOW predicts the target word based on the context. Consider the sentence: "The small cat drinks milk" as part of the learning corpus. If the word cat is the target word, then the words "the", "small", "drinks", and "milk" are the context words. The size-2 context corresponds to the two words before and after the target word one wants to embed. The context words are the input of the CBOW network, and the output is "cat". Skip-gram predicts the surrounding words given the current word. In the Skip-gram architecture, the target word "cat" is the input, and the context words are the output. The model tries to maximize the probability of predicting the surrounding words within a specified context window, a more complex task than learning the target word. Skip-gram is particularly efficient on rare words.

A similar network serves as the basis for both approaches in Word2Vec. Figure 3.2 represents, more specifically, a CBOW approach. The network

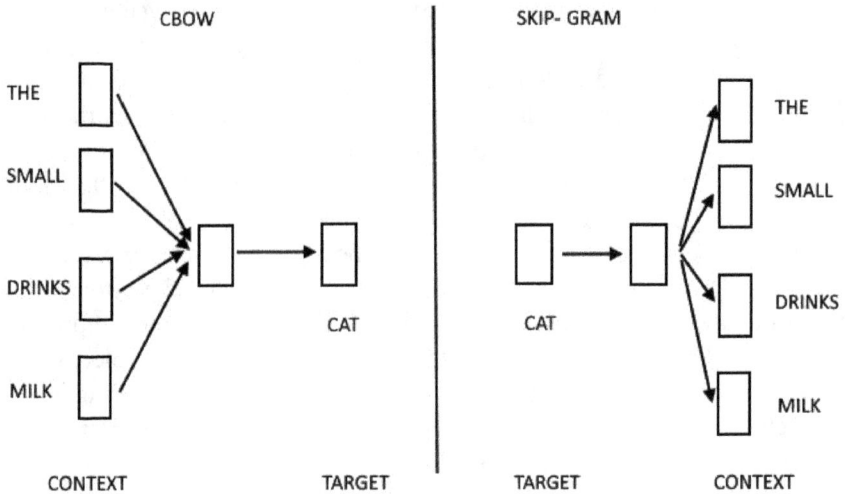

Fig. 3.1. The CBOWs predicts the current word based on the context. Skip-gram predicts the surrounding words given the current word.

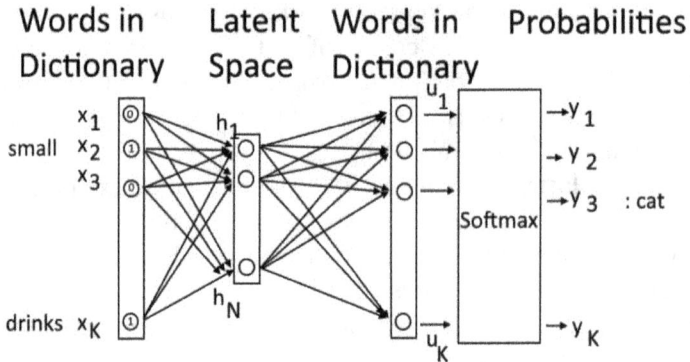

Fig. 3.2. The CBOWs architecture embeds the words in a dictionary by computing optimal matrices to predict the current word based on the context. Each arrow corresponds to a multiplication of the input by a number learned during training. The intermediary layer is the latent space, generally a low-dimension representation capturing the essential features of the higher-dimensional input space.

codes the possible associations of the words in the dictionary as an element of a matrix. In this example, the context words include the words around the word cat (Context window of size 1), namely "small" and "drinks". In CBOW, a value one is associated with each context word.

The values in the latent space are obtained through a linear transformation:

$$
\begin{pmatrix}
w_{1,1} & w_{1,2} & \cdots & w_{1,K} \\
\cdots & \cdots & \cdots & \cdots \\
w_{N,1} & \cdots & \cdots & w_{N,k}
\end{pmatrix}
\begin{pmatrix}
x_1 \\
x_2 \\
\cdots \\
x_K
\end{pmatrix}
=
\begin{pmatrix}
h_1 \\
\cdots \\
h_K
\end{pmatrix}.
\tag{3.1}
$$

Let us show how the matrix works on a single word. Let us assume that the first word in the dictionary is "small". One has

$$
\begin{pmatrix}
w_{1,1} & w_{1,2} & \cdots & w_{1,K} \\
\cdots & \cdots & \cdots & \cdots \\
w_{N,1} & \cdots & \cdots & w_{N,k}
\end{pmatrix}
\begin{pmatrix}
1 \\
0 \\
\cdots \\
0
\end{pmatrix}
=
\begin{pmatrix}
w_{1,1} \\
\cdots \\
w_{N,1}
\end{pmatrix}.
\tag{3.2}
$$

Embedding the word "small" corresponds to the first column of the matrix w. A second matrix, the output layer weight matrix, transforms the values in the latent space (hidden layer) into the word space. A softmax function scales the values in the output layer into a probability, for instance, for the word "cat". The softmax function transforms each value v_i in a vector, summing to one, using the function.

$$
f(v_i) = \frac{\exp(v_i)}{\sum_i \exp(v_i)}.
\tag{3.3}
$$

Learning aims to find the best embedding for each word and, therefore, the best elements for the embedding matrix and the output layer weight matrix. It is important to note that the word order in the context window does not affect the output word of the Word2Vec model. Typically, two or more words around the target word are processed.

Word2Vec permits comparing ancient text corpora, classifying them automatically according to their topics or styles, or studying their genealogy and reconstructing their transmission history (stemmatology). Following Word2Vec, several similar architectures were developed. For instance, an extension of Word2Vec to pages and paragraphs of variable length became a new state-of-the-art sentiment analysis at the time of publication (Le and Mikolov, 2014). The approach can also be extended to embed spoken words, sounds, or text segments into vectors. After embedding, the different vectors can be combined into a multimodal approach with various sources, one of the main approaches in recent AI tools.

The performance of computational approaches, as well as their scope, is steadily increasing. NLP supports complex tasks, such as intertextual studies. Intertextual connections represent allusions to other authors that are sometimes difficult to detect and are very important in literary analysis to understand authors' literary backgrounds. Burns *et al.* (2021) implement a Word2Vec approach for intertextual study centered on the texts of the Roman historian Livy. In other words, the task consists of finding allusions to Livy in the work of other authors and, therefore, understanding Livy's cultural influence.

The Word2Vec approach is also used in books, articles, or movie recommenders (Esmeli *et al.*, 2020). Word2Vec is often combined with another approach called TF-IDF (Wang and Shi, 2022), which stands for Term Frequency-Inverse Document Frequency. TF-IDF determines how important a word is within a document, considering the whole collection of documents (corpus). A frequent word in a specific document but less commonly used across the entire corpus has a high TF-IDF and may serve for keyword extraction or text summarization by identifying important sentences in the document.

An extension of the approaches to graphs, called Node2Vec, applies the Word2Vec approach, choosing nodes by randomly moving on a graph (Grover and Leskovec, 2016). Consider the graph in Fig. 3.3. The algorithm generates a series of paths by choosing the next node probabilistically. The original paper considers a random walk on the graph with possible steps to the nearest neighbor or a node two steps away. In the example of Fig. 3.3, the walk is at node 5. The next step may be a one-step move to node 4, 6, or 7 or a two-step move to node 2 or 3. The probability of returning to a previous node is very low to avoid generating sequences with repetitions. The unnormalized transition probabilities are proportional to the alpha factor of the considered step. The alpha factor to walk back to the previous node of the last transition is $\alpha = \dfrac{1}{p}$ with p a large constant. For a directly connected node (different from the previous node), one has $\alpha = 1$, and $\alpha = \dfrac{1}{q}$ for a two-step away walk ($p \gg q > 1$). Some sequences of some length are generated (i.e., 4,5,3,2,1) using the

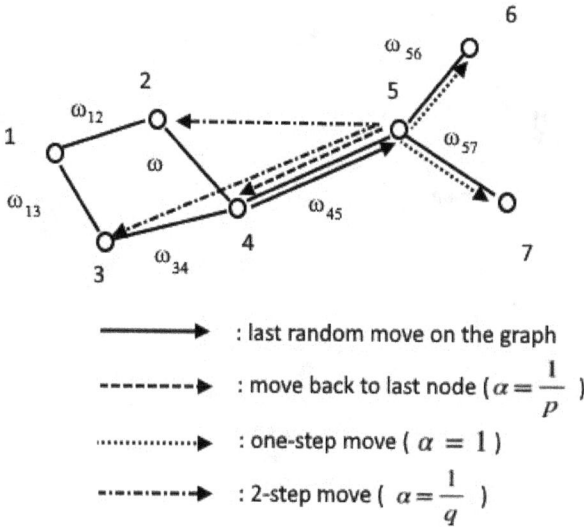

Fig. 3.3. The node2vec algorithm generates sequences on a graph following transition probabilities. The sequences are processed like text by a Word2Vec algorithm.

probabilities. The nodes are like words, and the sequence is the context. They are processed using an algorithm similar to that of Word2Vec.

Node2Vec is used routinely for social network analysis, recommendations (Palumbo *et al.*, 2018), or topic extraction (Zuo *et al.*, 2017). Given a graph representing the connections between the different films, actors, and topics completed by nodes representing user feedback and viewing history, Node2Vec defines proximity between a user, movies, and actors used to suggest new recommendations. Node2Vec can also be used to study a network where the nodes represent historical figures, and edges represent recorded interactions between them (letters, common meetings). The algorithm is also used to analyze ancient languages; Node2Vec analyzes glyph similarities between ancient Chinese characters' glyphs potentially related in origins or semantics (Chi *et al.*, 2022; Chen *et al.*, 2020).

Let us now discuss an example from mythology. We have analyzed a small corpus containing known instances in which the Big Dipper is identified as a group of men in mythology and folklore. Figure 3.4 shows the word embedding around the Big Dipper/Ursa Major using Skip-Gram with a context window of size 5. The embedding in two dimensions reflects some aspects of the story about seven men becoming the stars of the

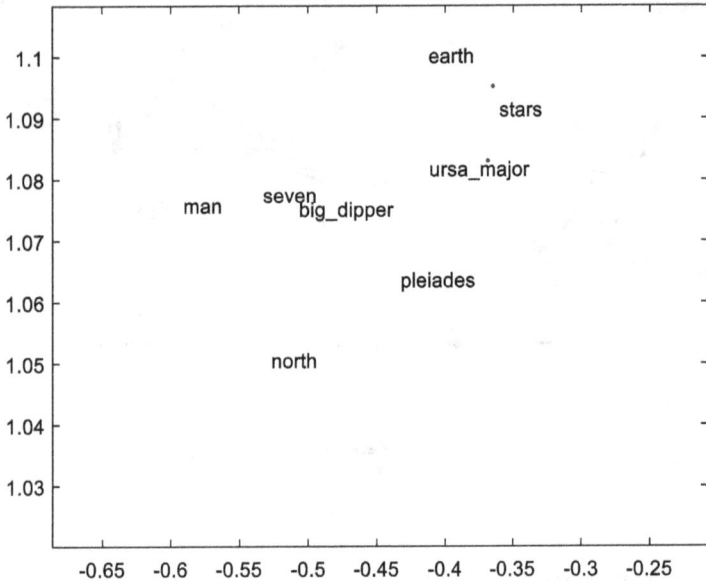

Fig. 3.4.   Word embedding of all versions of Starlores containing the motif of seven men as Ursa Major.

Big Dipper. In some stories in Eurasia, the seven men, the stars of the Big Dipper, are hunters who kidnap one of the Pleiades. Another cluster in the embedding (not shown) contains the words youngest, sister, bear, and dog. In America, several stories are about a young woman becoming a bear and killing everybody (Frank, 2023, 2024). A young woman's sister survives, protected by a dog, and becomes a star in the sky. The analysis using Word2Vec furnishes a reasonable embedding of the words, identifying two clusters of stories. In recent years, LLM has made a much finer analysis possible so that software tools can summarize the different versions of a story, identify differences, and group them reliably according to topics.

The Word2Vec algorithm was applied to an ancient Greek corpus (Köntges, 2020), the Nordic saga of the Younger Edda from the 10th century, and Celtic tales (Fumanal-Idocin *et al.*, 2023). A strong association was found between Zeus and Herakles. The analysis goes one step further by analyzing the affinities between the main characters by the weight of their connections (The number of common occurrences of the two characters in the corpus using a window). Freya has, for instance, a high affinity with Loki, possibly created by the main topic of death. List (2022) applies Word2Vec to Koine Greek (circa 2nd century BCE to 2nd century CE) to discover prototypical, culturally salient associations of words in Koine Greek.

Let us conclude this chapter with some words on a study of the evolution of the story of Gilgamesh. Sentiment analysis extracts mood or sentiments (love, hate, fear) in text or videos (Alslaity and Orji, 2024). The evolution of the characters' psychology in the "Epic of Gilgamesh" is based on its ancient versions, which have spread over 2000 years, with the last text used in that study being from 100 BC. The analysis by Du *et al.* (2023) goes one step further by analyzing the evolution of the main character's psychological description over two millennia. The analysis matches the adjectives and verbs describing the characters to contemporary psychological trait dictionaries. By comparing the evolution of the terms through time, the authors track the changes in the description of the character's psychology using multi-dimension embedding (MDS, see Chapter 1) on the matrix of co-occurrences between words. Extraversion is the most frequent personality trait in Sumerian, Old Babylonian, and Middle Babylonian texts. Agreeableness is also a frequent trait dimension. Both intellect/openness and conscientiousness are important only in the late versions. The analysis shows a continuous shift toward contemporary personality models in later periods. Considering that we do not have access to the psychology of people so far in the past, this study opens a new window into using NLP to get a glimpse of the past values and psychologies conveyed in epic stories and narratives. Using more advanced approaches in future work will probably give a finer analysis of the characters. NLP has progressed at a tremendous pace in the last few years. Embedding is central to any NLP technique and a central element in the methods presented in the next chapter.

Nowadays, Word2Vec is still used for word embedding as a starting point for complex text analysis. Word2Vec is an important method used in NLP, but not the only one. Studies consistently show the superiority of the best LLM over classical approaches (Uthirapathy and Sandanam, 2023) for topic modeling or sentiment analysis. Word2Vec is still important in introducing modern ideas in LLMs, which we will discuss in the next section. In particular, the "attention" mechanism, the main building block in LLMs, is easier to understand once the principles behind embedding have been explained.

3.3 Recurrent Neural Networks and Long Short-Term Memory

Figure 3.5 sketches the two main categories of networks: the feedforward network, in which input data are processed through a series of transformations, and the recurrent network, in which information is processed on loops in several iteration steps.

In a feedforward network, a cell processes the information using a linear relation:

$$Y = \sum_k W_k X$$

where W_k is a matrix and, therefore, a linear transformation of the elements in X. A nonlinear transformation (for instance, a ReLu) is sequentially added to the linear part.

The recurrent neural networks (RNNs) made processing (ordered) word sequences possible. They process whole sentences by having an architecture in which the output from the previous step is fed back as input for the current step. Taking the sentence "The cat eats the . . . ", the recurrent system processes the words "cat", "eats", and "the" sequentially and may suggest the word mouse or bird.

In a recurrent model, some hidden states of the neural cell at step t are memorized and combined with the next input. Contrary to a feedforward network in which the information is processed through the complete network at each iteration, the recurrent network "re-injects" information h from the past iteration steps into the current step.

LSTM, a special type of recurrent network (Fig. 3.6), received much attention as it first solved the major problem with deep learning, the

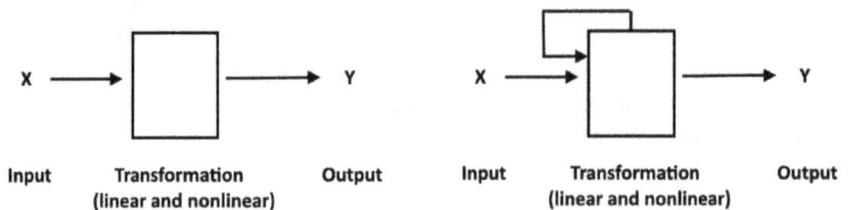

| Input | Transformation (linear and nonlinear) | Output | Input | Transformation (linear and nonlinear) | Output |

Fig. 3.5. (Left) A feedforward network cascades units in which the information is processed from left to right in one step. (Right) A recurrent network processes the data sequentially, supplying the next input and the internal states to the next processing step.

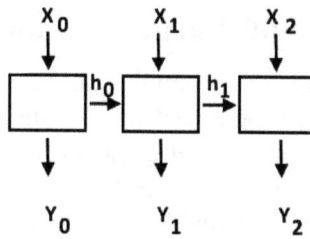

Fig. 3.6. Recurrent LSTM structure to process, for instance, a sentence. The output Y2 corresponds to the next word and depends on the input data and memory state of the previous steps.

so-called vanishing gradient problem introduced in Chapter 2. Neural networks are usually trained using a backpropagation mechanism or another standard gradient approach. The network will not learn well if the gradients are too small or very noisy, which is generally true with deep learning with its many layers without some signal re-injection mechanism. Hochreiter and Schmidhuber (1997) suggested networks that solved that issue by adding a memory cell that "optimally" stores the information during learning; the network is called, for that reason, a Long Short-Term Memory (LSTM). While LSTMs are still very performant for some tasks, they do not scale up well to large sentences, requiring much memory and complex tuning.

Recurrent networks have applications for ancient text recovery. Fetaya *et al.* (2020) implement RNNs to assist researchers in completing some missing information. Babylonian texts are often on damaged clay tablets, and much information is missing. With an LSTM, the model processes the Late Babylonian period's highly structured legal, economic, and administrative texts. The network makes proposals for the missing signs. In another study, Papavassileiou *et al.* (2023) applied a bidirectional recurrent network architecture to complete missing parts in Mycenaean texts. As the number of texts for learning is scarce, the data are augmented to have larger sequences. For example, the authors swap the word referring to the toponyms or the owner's name in some documents, generating a larger corpus.

RNNs furnished some of the best results in completing a missing ancient text and deciphering text in (known) languages (Ferrara and Tamburini, 2022).

3.4 Transformers and Large Language Models

LLMs like the one in ChatGPT, Gemini AI, or Claude can predict the next word in a sentence from previous sentences. In the example of ChatGPT3, the model learned 175 billion weights, essentially numbers packed in large matrices that process the input to suggest the best next word. GPT3 embeds words in 12,228 dimensions, meaning that over 12,000 numbers represent each token. The newest LLM models are still much larger. Nowadays, a token may be a word, a node, some image, or a sound to give examples. Training the system from scratch is hugely demanding in computing power and, therefore, only accessible to private and academic researchers with access to powerful machines. After learning, the system represents each possible token as a vector. A value in the embedding vector may be associated with some property (cat and lion may have a value almost identical to the word fur or another with a value close to the one for feline), making the embedding somewhat interpretable. However, it is more of an exception than a general rule that a single value can be univocally associated with a property. The attention or self-attention mechanism modifies the vector corresponding to a word to capture its relationships to a context. The attention mechanism is described in Section 3.4.2. The context can be very large nowadays and may include several hundred pages. It is much more than the context used in Word2Vec, which typically considers only a few words around the target word.

The similarity between the embedding of two words tells us how close the two words are. The word cat is close to feline or lion. A way to measure the proximity of two words is to use the dot product, which consists of multiplying each element in the embedding vector of one word with each component of the second vector. Given a 3D embedding of cat = (0.3, 0.5, 0.7) and lion = (0.4, 0.5, 0.6), the dot product is 0.3 * 0.4 + 0.5 * 0.5 + 0.7 * 0.6 = 0.78. The dot product has high values if both embeddings represent a vector pointing in the same direction and is negative if they point to opposite directions.

3.4.1 *Positional encoding*

The encoding of words contains two types of information: word embedding and positional encoding. Positional encoding is intended to be a small correction to word embedding and, therefore, must stay small. Position encoding is one of the main components of LLM. Given the incomplete sentence, "the cat is a loved...", the system learns better which word to follow if the different tokens' probable positions are learned during training. The sine and cosine functions were originally used as their values are bounded between one and minus one. The functions permit uniquely coding a word's position relative to the others.

3.4.2 *Attention is all you need!*

Vaswani *et al.* (2017) broadly introduced a new approach to solving the main issue with NLP: the difficulty of understanding the large context around words. The context may depend on words quite far away from the considered sequence in a sentence. The context modifies the choice of the next word. Consider the two sentences: "I sit on the river's bank" and "I go to the bank." Self-attention aims to disambiguate the word bank by analyzing its context to choose the best next word. Self-attention can be understood within the context of word embedding. In Word2Vec, the embedding is static. After training on some corpus, each word has a fixed vector representation independent of its context within a sentence. Word2Vec captures semantic relationships between words in the considered corpus. In self-attention, word embedding depends on the words around it, its contextual embedding. A second difference is that the word position is also encoded in a matrix.

The original paper was directed toward translation; therefore, the architecture used an encoder and a decoder (see Fig. 3.7). The encoder modifies the word embedding with several processing layers using each the self-attention mechanism, a normalization block to ensure values are within some ranges, a feedforward network, and some shortcut paths, as in ResNet. As discussed in the previous chapter, nonlinearities are essential to learning in neural networks. The nonlinearity generally uses a ReLu function (see complement Chapter 2). The decoder furnishes the next word in the target language. It uses blocks with a similar structure to the encoder.

One of the main novelties of Vaswani *et al.* (2017) was the intensive use of the attention mechanism. Attention is central to the recent development of NLP. Its function is to modify the embedding of a word depending on the context using three matrices: Q, K, and V. The three matrices Q, K, and V are called the query, key, and value matrices. They modify the embedding depending on the context of the words close to the target word. Consider the beginning of a sentence, "the big cat". A word embedding is associated with each word. In the attention algorithm, the embedding is modified by adding the values of the position encoding for the three corresponding words.

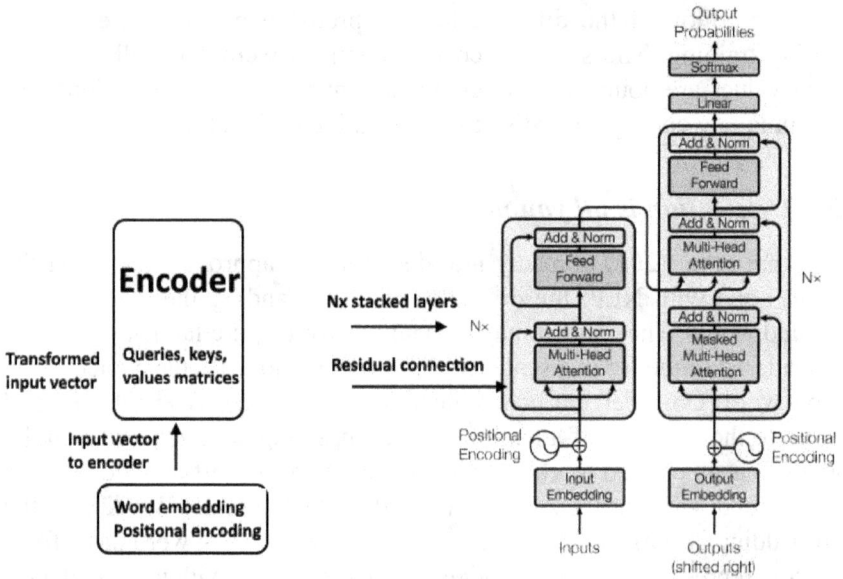

Fig. 3.7. Transformer architecture using the self-attention mechanism. Modified from Vaswani *et al.* (2017), CC-BY license. The left part focuses on the encoder.

As the values for positional encoding are bounded, they shift a word by some small value depending on the order of the words in a sentence.

The input embeddings are modified by multiplying it with three (learned) weight matrices: Wq, Wk, and Wv. The values matrices Wq, Wk, and Wv result from the training on a very large corpus and are the same regardless of the specific words in the input sequence to obtain the query Q, key K, and value matrices V. Figure 3.8 shows an example for the key matrix.

The context matrix is obtained by multiplying the query and the key matrices (the words query and key have no deep meaning; they were used in the original paper and stayed). The context matrix represents the similarity or relevance between two words in the input sequence. The higher the value, the more relevant the key is to the query:

$$Q\,K^T = \begin{pmatrix} \cdot & \cdot & \cdot \\ \cdot & \cdot & \cdot \\ \cdot & \cdot & \cdot \end{pmatrix} \begin{matrix} \text{'The'} \\ \text{'big'} \\ \text{'cat'.} \end{matrix}$$

Weights   Word   Position
Embedding   Embedding

$$K^T = \begin{pmatrix} w_{1,1} & w_{1,2} & \cdots & w_{1,d} \\ w_{2,1} & w_{2,2} & \cdots & w_{2,d} \\ \cdots & \cdots & \cdots & \cdots \\ w_{d,1} & w_{d,2} & \cdots & w_{d,d} \end{pmatrix} \left(\begin{pmatrix} E_{1,the} & E_{1,cat} & E_{1,is} \\ E_{2,the} & E_{2,cat} & E_{2,is} \\ \cdots & \cdots & \cdots \\ E_{d,the} & E_{d,cat} & E_{d,is} \end{pmatrix} + \begin{pmatrix} P_{1,the} & P_{1,cat} & P_{1,is} \\ P_{2,the} & P_{2,cat} & P_{2,is} \\ \cdots & \cdots & \cdots \\ P_{d,the} & P_{d,cat} & P_{d,is} \end{pmatrix} \right)$$

Fig. 3.8. A (trained) weight matrix W transforms the embedding and position vectors into the key matrix. The query Q and value V matrices are similarly computed.

The T state for transpose corresponds to reversing lines and columns in the matrix. In the example, each element in the matrix furnishes a value related to the similarity between words ("the", "big", "cat") within the input sequence. Let us recall that linear transformations may represent geometric transformations, including shearing, dilation, and translations. One may intuitively understand $Q K^T$ as a context matrix relating the semantic and positional relationships in a transformed space.

One can picture the context matrix as an indication of the similarities of the words in the window, transforming word embedding depending on the context. The attention is defined below:

$$Attention(QKV) = softmax\left(\frac{QK^T}{\sqrt{d}} \right) V, \qquad (3.4)$$

using Eq. (3.3) on each row for the softmax. The matrix Q and K have the same dimension ($N \times d$) with as many rows as given words (or, more generally, tokens) in the input sequence and columns of size equal to the word embedding dimension d. It follows that the matrix $Q K^T$ is square ($N \times N$) with N the size of the input sequence ($N = 3$ in the example of the input sequence: the big cat). The matrix V contains numerical representations of the embedded words ($N \times d$ matrix). After learning, the cascade of attention and feedforward blocks permits the determination of the next word.

What we have just described is a single-head attention. The system runs several parallel attention blocks in a transformer, each with Q, K, and

V values. Each block captures other relationships in the input sequence, such as sentiment analysis or style. One speaks of multi-head attention.

Finding the next word in a translation is the task of the decoder. Let us assume that the encoder has processed the input sentence, the big cat, and the output is in French *le gros* .... In the final step, the decoder furnishes a probability for each word in the dictionary to be the next word and generates the next word in the translation, probably the word "chat". Advanced LLMs generate several answers in parallel and delay their output to choose the best answer.

Transformers were successfully adapted to process sound, images, or videos. An image is split into patches in a Visual transformer (ViT) (Dosovitskiy, 2020). Each image patch is transformed into a vector by aligning the pixels, a simple operation. A linear projection layer maps the vector to a lower-dimensional embedding space. Each patch's vector representation (embedding) is input to a transformer. Image patches are treated like tokens (words) in an NLP application.

3.5 Applications of Transformers to Ancient Languages

The success of BERT in English and many modern languages is starting to impact the study of ancient languages. While computational linguistics already has a long story with many performant software tools for Latin and Ancient Greek, for instance, the Classical Language Toolkit working on 19 ancient languages (Johnson *et al.*, 2021), one observes the emergence of neural networks for tasks that, till recently, mostly used fixed rules and lexica for analysis. Transformer models are designed to capture the relationships between words in a sentence. They can be adapted for standard tasks in linguistics, such as part-of-speech analysis (noun, verb), dependency parsing (a tree-like representation of the grammatical relationships and dependencies between words in a sentence), or name entity recognition (Cesar, legion), lemmatization (i.e., running → run). For instance, Riemenschneider and Frank (2023) used an encoder-decoder transformer architecture for lemmatization (Raffel *et al.*, 2020). After training, the system output is the lemmatized form of a word.

Sommerschield *et al.* (2023) survey ML in Ancient Languages, proposing the taxonomy: Digitalization, Restoration, Attribution, Linguistic Analysis, Textual Criticism, and Translation. CNN approaches are mostly for glyph analysis, digitization, and quality enhancement. There are primarily computer vision tasks. Textual restoration, attribution, linguistic analysis, and sentiment analysis are the domains of transformers and RNNs. BERT and GPT are the dominant neural networks for text analysis of ancient languages.

A Latin-BERT built by Bamman and Burns (2020) uses Latin sources from 200 BCE to the present day with over 600 million words. Latin-BERT is state of the art for part-of-speech tagging, which is the task of assigning a part-of-speech label (e.g., "noun", "verb") to each word in a text. The system outperforms the standard word embedding approaches like Word2Vec and its derived algorithms. The performances of the best generalist LLM systems are quite remarkable. Summaries of 16th century letters in Latin and early New High German furnish versions of qualities comparable to those of human experts (Volk *et al.*, 2024).

3.5.1 *Handwriting recognition and decipherment*

Transformers have an advantage over CNN as they can integrate different sources and foster the development of multimodal systems. Combining a Vision Transformer (Bao *et al.*, 2021) as an encoder with a language decoder permits the transformation of an image into some text. Li *et al.* (2021) trained such a system to recognize handwriting and translate the text into English. Ströbel *et al.* (2022) tested the system on 16th century correspondences in Latin and found that it outperformed the state-of-the-art tools.

Decipherment is an interesting testing bench for neural networks. Snyder *et al.* (2010) succeeded in deciphering Ugaritic using Bayesian analysis with some success. The method uses the proximity between Ugaritic and Hebrew. As Ugaritic has been known for some time, results validation is straightforward. Luo *et al.* (2019, 2021) could improve the results by combining two LSTM networks, one per language related by an attention mechanism. The same approach was also successfully applied to Linear B and ancient Greek, two close languages, but with the supplementary difficulty of Linear B being a syllabic writing system.

3.5.2 *Textual restoration*

Transformer-based approaches are often preferred to CNN for text restoration of ancient Greek or Latin texts. CNN is often more appropriate for glyphs or logograms (Chi *et al.*, 2022; Kang *et al.*, 2021). Many ancient Greek inscriptions have survived but are partially damaged, and virtual restoration tries to infer the missing text. The Ithaca system is available on the internet with an intuitive interface to restore ancient Greek texts and propose a date and a region of origin (Assael *et al.*, 2022). The system suggests several possible outputs with their probabilities. Interestingly, the authors report that the system is often less performant than the domain specialists but boosts their performance when they use it as a recommender. This observation emphasizes again the role of AI as an assistant.

3.5.3 *Attribution*

State-of-the-art is less advanced for ancient Greek languages than Latin. Studies generally use classical approaches (Word2Vec), a modern Greek BERT, or fine-tuned the open source BERT. The network is fine-tuned by retraining the system to improve it on a particular task (Hoang *et al.*, 2019). Yamshchikov *et al.* (2022) applied BERT models to the Plutarch of Chaeronea surviving corpus (ca. 45–120 CE). The corpus contains several texts attributed to other authors. The study tried to clarify the different origins of the texts. The limited size of the Plutarch corpus does not allow the development of a system from scratch, and learning is done by fine-tuning BERT with English or modern Greek translations. The resulting model was specifically fine-tuned for authorship attribution in Ancient Greek. The system got 80% accuracy on 18 authors' attribution on 500 lines per author set aside during learning. The network recognizes its region of origin with about the same accuracy as for authorship. The authors considered three large areas: Delphi (Plutarch's home for many years), a region around Alexandria, and the central part of the western coast of modern Anatolia. The study found similarities between two texts from the Pseudo Plutarch and second- and third-century CE texts by authors associated with Alexandria's region.

3.5.4 *Topic and sentiment analysis*

Transformers have been adapted to multiple tasks. Sentence BERT or S-BERT is a clever method for sentence embedding. Let us recall that word embedding attributes a vector of numerical values to each word, depending on the context. In a first approximation, a sentence embedding may be computed by averaging the embedding of each word in a sentence with a transformer. The embedding can be refined using a Siamese or Triplet network. In Chapter 2, we mentioned the usefulness of Siamese networks in quantifying distances by processing two images with the same network with identical weights. A distance can be rationally computed by comparing their values in a latent layer. A similar approach was developed for sentence analysis using a Siamese BERT model. The model learns from examples to recognize close sentences. After learning, the model can compute the distance between two sentences. Reimers and Gurevych (2019) implement triplet networks to discover sentence similarities and differences. During the learning phase, the network receives an anchor sentence, an almost identical sentence, and a different sentence to learn how to classify the distance between sentences and meaningful sentence embeddings. The training proceeds on libraries of sentences, such as the Multi-Genre Natural Language Inference (MultiNLI) corpus, with 433k examples, created by crowd workers (Williams *et al.*, 2018) who received the task to write for each sentence taken from multiple sources a correct sentence, one that might be correct and one that is definitively incorrect.

After sentence embedding, a dimension reduction approach (see Chapter 1) can be applied to visualize possible groupings. A TF-IDF algorithm permits the clusters to be tagged with the main topics as words or sentences.

Tangherlini and Chen (2024) analyzed Hans Christian Anderson's fairy tales and travel writings with S-BERT and identified intertextual interdependencies across genres and time.

Chandra and Ranjan (2022) applied S-BERT to the Upanishads (a collection of late Vedic texts) and the Bhagavad Gita. to embed the sentences, which were then processed by a topic detection tool. They found a very high similarity between the topics of these two texts, which is an expected result. Quantifying the similarity using a measure, they found that of the fourteen topics extracted from the Bhagavad Gita, nine are similar to the topics of the Upanishads by more than 70%.

Sentiment analysis is one of the main commercial applications used to understand a group of people's sentiments toward a topic or a company. Sentiment analysis is a very useful tool for automatically capturing some shifts in the emotional state of the main characters and writers. Chandra and Kulkani (2022) use a sentiment analysis on versions of the Bhagavad Gita, a sacred text in Hinduism, originally written in Sanskrit and translated into English. The Gita is now part of the Mahabharata, one of the oldest and longest-known epics. Using a hand-labeled Twitter-based training dataset that captures 10 sentiments, they fine-tuned BERT specifically for sentiment analysis. Their study found that although the style and vocabulary in the respective translations widely vary, the different translations captured mostly the same sentiments. The statistical analysis of a very large text body may furnish good starting points for further studies. On the negative side, the classification of ten sentiment categories obtained from contemporary tweets is an oversimplification of the analysis of the complex emotions conveyed by ancient texts.

3.5.5 *Outlook*

As emphasized by Sommerschield *et al.* (2023), "progress in ML relies not only on powerful models, but also on the quality and quantity of datasets, evaluation metrics, and experimental protocols." Model performance depends very much on the quality of the data. Learning is way better on a correct and well-preprocessed data set. The rule holds for any ML problem!

Hallucinations and, more generally, the validation of results are huge issues in LLM. Different strategies improve LLMs. For instance, systems generate better answers if the context includes the prompts and some retrieved information from external knowledge sources (Retrieval Augmented Generation, RAG). Some AI agents may also help retrieve helpful information. While much work is done on that topic, some of the latest developments are not public domain (Huang *et al.*, 2024).

The most advanced NLP systems offer a new and fresh approach to analyzing ancient texts. Neural networks provide powerful methods to address problems where researchers may otherwise be overwhelmed with information. LLMs generally require a large corpus for learning, which is still a problem for many ancient languages. Overfitting is a major issue if the neural network contains many parameters compared to the number of words used for learning. The model loses its ability to work well on new, unseen data. The difficulty is not specific to Ancient Languages but extends to any field with a small set of learning and testing data (Calder *et al.*, 2022).

Besides Latin and Greek, several ancient languages have a large corpus to train a network. India and China have a very large collection of digitalized ancient writings, for instance, the corpus of Siku Quanshiu (Liu *et al.*, 2023) or the Vedas with versions in Sanskrit using Unicode to facilitate processing (Panday *et al.*, 2024). Cuneiform writing has also a very large corpus of texts in different languages. Semitic Akkadian is, for instance, an ancient language with a large cuneiform corpus. Digitalizing the corpus in a form suitable to NLP is still a big challenge (Smidt *et al.*, 2024). The dialog between experts and data scientists will also permit refining the expert knowledge provided to the neural network and lead to more subtle analysis.

Advanced ML systems are increasingly capable of deciphering languages, potentially leading to breakthroughs in our understanding of ancient and unknown scripts. These systems will produce a mixture of accurate and inaccurate or misleading results. Human expertise remains crucial for validating the findings and separating genuine discoveries from errors.

The answer' quality of the best LLMs is quite surprising, and much research is underway to understand the emergent properties of LLMs, not least for safety issues. The studies by Reddy (2023) and Kadkhodaie *et al.* (2023), for instance, demonstrate that learning depends on how the data are presented during learning or on the details of the networks. A central question in AI is whether the models have learned so many examples that they can access a lot of stored knowledge or if neural networks also show some emergent properties. Finding good measures of the network's generalization capabilities is an important general research topic in AI (Yang *et al.*, 2023). Experiments have shown that networks can sometimes learn an underlying structure without being explicitly trained. Many instances of generalizations in LLM are reported (Dong, 2022). Generalization capabilities often appear suddenly after some slow learning phase, similar to a phase transition in physics, like between water and ice. Let us discuss two applications showing such a transition.

In-context learning is a much-studied instance of LLM's emergent behavior. In-context learning is the capacity to learn from a few examples the user provides without additional weight updates (Olsson *et al.*, 2022). The user may prompt a few sentences expressing sentiments about movies. When prompted with a new movie title, the LLM automatically associates some sentiment about the movie without being asked for. The basic mechanisms behind in-context learning are believed to be similar to the ones making LLM so efficient at writing coherently. Attention blocks are at the core of LLM's. Transformers use multiple attention heads grouped in blocks. An attention head processes the data very differently from another one despite having the same structure. An attention head may, for instance, furnish the last but one word (or item) as output or associate a sequence of three words (or items) that may appear separated in the context: the big cat eats the brown mice (cat eats mice in the example). Combining two attention heads is the minimal building block allowing in-context learning (Reddy, 2023). The two attention head generalization capabilities undergo a sharp transition during training.

LLM are not the only neural networks underlying a transition in learning. The capacities of latent diffusion networks (Section 2.5.1) to generate a human face also undergo a rapid transition from a network capable of reproducing many faces to a network capable of generating new, very realistic faces. The experimental results of neural networks for face generation surprised the mathematical community as the very high problem dimension should normally require a much larger network if not specially optimized. Kadkhodaie *et al.* (2023) have shown that the network structure is particularly well suited to the problem of denoising faces and is sub-optimal for other problems. In other words, the network implicitly uses optimized functions to denoise face images.

The two examples prove that some neural networks have emergent behaviors that enhance the computational intelligence of the system after some learning. A better understanding of these emergent properties may lead to better reasoning performances of LLM.

LLMs are increasingly capable of generating software code and solving simple problems by answering questions in programming languages. This ability can bridge the gap between Humanities and Digital Humanities by reducing the initial investment needed to learn and use software for research and analysis, provided that robust quality assurance and verification processes are in place.

Transformer-based tools represent a quantum leap in computational linguistics, even if they have long struggled with simple tasks like counting the number of "r" in strawberry, a simple task for a rule-based algorithm! We observe a trend toward hybrid models that combine transformer-based or, more generally, neural networks with rule-based or chain-of-thought methods (Wei *et al.*, 2022). These models will extend the application range of neural networks in studying ancient languages. Multimodal systems integrating text and images may lead one day to new exciting applications, like the automatic identification of mythological characters on paintings and vases and relating them to ancient texts (Goutam *et al.*, 2024). AI may help preserve ancient dying languages for future generations by coding their structure in an LLM and many stories in a database (Mohanty *et al.*, 2024).

Benchmarks show that AI tools outperform standard approaches (Riemenschneider and Frank, 2023) in several standard computational linguistic tasks. Results obtained through AI complement, enrich, or sometimes confuse experts, but AI will certainly find a place in the Humanities. With time, the strengths and weaknesses of AI tools will be better understood, leading to better results. Even if the number of AI applications for ancient languages is rapidly expanding, some of the promises of AI will probably need time to be fulfilled, and many questions will stay outside of the reach of AI.

References

Alslaity, A., & Orji, R. (2024). Machine learning techniques for emotion detection and sentiment analysis: Current state, challenges, and future directions, *Behaviour & Information Technology*, 43(1), pp. 139–164.

Assael, Y., Sommerschield, T., Shillingford, B., Bordbar, M., Pavlopoulos, J., Chatzipanagiotou, M., ... & de Freitas, N. (2022). Restoring and attributing ancient texts using deep neural networks, *Nature*, 603(7900), pp. 280–283.

Bamman, D., & Burns, P. J. (2020). Latin Bert: A contextual language model for classical philology, arXiv preprint arXiv:2009.10053.

Bao, H., Dong, L., Piao, S., & Wei, F. (2021). Beit: Bert pre-training of image transformers, arXiv preprint arXiv:2106.08254.

Burns, P. J., Brofos, J. A., Li, K., Chaudhuri, P., & Dexter, J. P. (2021). Profiling of intertextuality in Latin literature using word embeddings, *Proceedings 2021 Conference of the North American Chapter of the Association for Computational Linguistics: Human Language Technologies*, pp. 4900–4907.

Calder, J., Coil, R., Melton, J. A., Olver, P. J., Tostevin, G., & Yezzi-Woodley, K. (2022). Use and misuse of machine learning in anthropology, *IEEE BITS the Information Theory Magazine*, 2(1), pp. 102–115.

Chandra, R., & Kulkarni, V. (2022). Semantic and sentiment analysis of selected Bhagavad Gita translations using BERT-based language framework, *IEEE Access*, 10, pp. 21291–21315.

Chandra, R., & Ranjan, M. (2022). Artificial intelligence for topic modelling in Hindu philosophy: Mapping themes between the Upanishads and the Bhagavad Gita, *PLoS One*, 17(9), p. e0273476.

Chen, H. Y., Yu, S. H., & Lin, S. D. (2020). Glyph2Vec: Learning Chinese out-of-vocabulary word embedding from glyphs. *Proceedings 58th Annual Meeting of the Association for Computational Linguistics*, pp. 2865–2871.

Chi, Y., Giunchiglia, F., Shi, D., Diao, X., Li, C., & Xu, H. (2022). ZiNet: Linking Chinese characters spanning three thousand years, *Findings of the Association for Computational Linguistics: ACL 2022,* pp. 3061–3070.

Devlin, J., Chang, M. W., Lee, K., & Toutanova, K. (2018). Bert: Pre-training of deep bidirectional transformers for language understanding, arXiv preprint arXiv:1810.04805.

Dong, Q., Li, L., Dai, D., Zheng, C., Wu, Z., Chang, B., ... & Sui, Z. (2022). A survey on in-context learning, arXiv preprint arXiv:2301.00234.

Dosovitskiy, A. (2020). An image is worth 16x16 words: Transformers for image recognition at scale, arXiv preprint arXiv:2010.11929.

Du, A. H., Karl, J. A., Fetvadjiev, V., Luczak-Roesch, M., Pirngruber, R., & Fischer, R. (2024). Tracing the evolution of personality cognition in early human civilizations: A computational analysis of the Gilgamesh epic, *European Journal of Personality,* 38(2), pp. 274–290.

Esmeli, R., Bader-El-Den, M., & Abdullahi, H. (2020, July). Using Word2Vec recommendation for improved purchase prediction, *2020 International Joint Conference on Neural Networks (IJCNN), (IEEE),* pp. 1–8.

Ferrara, S., & Tamburini, F. (2022). Advanced techniques for the decipherment of ancient scripts, *Lingue e Linguaggio,* 21(2), pp. 239–259.

Fetaya, E., Lifshitz, Y., Aaron, E., & Gordin, S. (2020). Restoration of fragmentary Babylonian texts using recurrent neural networks, *Proceedings of the National Academy of Sciences,* 117(37), pp. 22743–22751.

Frank, R. M. (2023). The European Bear's Son tale: Its reception and influence on indigenous oral traditions in North America, *Folklore: Electronic Journal of Folklore,* (88), pp. 119–146.

Frank, R. M. (2024). *The Skylore of an Indigenous People: The Case of the Performance Art and Traditional Beliefs of the Lenape Delaware (Algonquian) of the Northeastern Woodlands.* Open book.

Fumanal-Idocin, J., Cordón, O., Dimuro, G. P., López-de-Hierro, A. F. R., & Bustince, H. (2023). Quantifying external information in social network analysis: An application to comparative mythology, *IEEE Transactions on Cybernetics.*

Goutam, A., Trivedi, S., Goudar, V., Pandey, N., & Barve, S. S. (2024). Mythological story teller: Extending scope of deep learning to Indian mythology, *Nanotechnology Perceptions,* pp. 1287–1304.

Grover, A., & Leskovec, J. (2016). node2vec: Scalable feature learning for networks, *Proceedings 22nd ACM SIGKDD International Conference on Knowledge Discovery and Data Mining,* pp. 855–864.

Heggarty, P., Anderson, C., Scarborough, M., King, B., Bouckaert, R., Jocz, L., ... & Gray, R. D. (2023). Language trees with sampled ancestors support a hybrid model for the origin of Indo-European languages. *Science,* 381(6656), p. eabg0818.

Hoang, M., Bihorac, O. A., & Rouces, J. (2019). Aspect-based sentiment analysis using Bert, *Proceedings 22nd Nordic Conferences on Computational Linguistics*, pp. 187–196.

Hochreiter, S., & Schmidhuber, J. (1997). Long short-term memory, *Neural Computation*, 9(8), pp. 1735–1780.

Huang, X., Ruan, W., Huang, W., Jin, G., Dong, Y., Wu, C., ... & Mustafa, M. A. (2024). A survey of safety and trustworthiness of large language models through the lens of verification and validation, *Artificial Intelligence Review*, 57(7), p. 175.

Johnson, K. P., Burns, P. J., Stewart, J., Cook, T., Besnier, C., & Mattingly, W. J. (2021). The classical language toolkit: An NLP framework for pre-modern languages. *Proceedings 59th Annual Meeting of the Association for Computational Linguistics and the International Joint Conference on Natural Language Processing: System Demonstrations*, pp. 20–29.

Kadkhodaie, Z., Guth, F., Simoncelli, E. P., & Mallat, S. (2023). Generalization in diffusion models arises from geometry-adaptive harmonic representation, arXiv preprint arXiv:2310.02557.

Kang, K., Jin, K., Yang, S., Jang, S., Choo, J., & Kim, Y. (2021). Restoring and mining the records of the Joseon dynasty via neural language modeling and machine translation, arXiv preprint arXiv:2104.05964.

Köntges, T. (2020). Measuring philosophy in the first thousand years of Greek literature, *Digital Classics Online*, pp. 1–23.

Le, Q., & Mikolov, T. (2014). Distributed representations of sentences and documents, *International Joint Conference on Machine Learning*, (PMLR), pp. 1188–1196.

Li, M., Lv, T., Chen, J., Cui, L., Lu, Y., Florencio, D., ... & Wei, F. (2023, June). Trocr: Transformer-based optical character recognition with pre-trained models, *Proceedings of the AAAI Conference on Artificial Intelligence*, 37(11), pp. 13094–13102.

List, N. (2022). How can we investigate ancient Greek categories without the influence of our own? Exploring kinship terminology using word2vec, *International Journal of Lexicography*, 35(2), pp. 137–152.

Liu, C., Wang, D., Zhao, Z., Hu, D., Wu, M., Lin, L., ... & Zhao, L. (2024). SikuGPT: A generative pre-trained model for intelligent information processing of ancient texts from the perspective of digital humanities. *ACM Journal on Computing and Cultural Heritage*, 17(4), pp. 1–17.

Luo, J., Cao, Y., & Barzilay, R. (2019). Neural decipherment via minimum-cost flow: From Ugaritic to Linear B, *Proceedings 57th Annual Meeting Association for Computational Linguistics*. Florence, Italy, (Association for Computational Linguistics), pp. 3146–3155.

Luo, J., Hartmann, F., Santus, E., Barzilay R., & Cao, Y. (2021). Deciphering under segmented ancient scripts using phonetic prior. *Transactions of Association for Computational Linguistics* 9, pp. 69–81.

Mikolov, T., Sutskever, I., Chen, K., Corrado, G. S., & Dean, J. (2013). Distributed representations of words and phrases and their compositionality, *Advances in Neural Information Processing Systems*, 26, pp. 1–9.

Mohanty, S. S., Dash, S. R., & Parida, S. (eds.) (2024). *Applying AI-based Tools and Technologies Towards Revitalization of Indigenous and Endangered Languages*, (Springer).

Olsson, C., Elhage, N., Nanda, N., Joseph, N., DasSarma, N., Henighan, T., ... & Olah, C. (2022). In-context learning and induction heads, arXiv preprint arXiv:2209.11895.

Palumbo, E., Rizzo, G., Troncy, R., Baralis, E., Osella, M., & Ferro, E. (2018). Knowledge graph embeddings with node2vec for item recommendation, *The Semantic Web: ESWC 2018 Satellite Events: ESWC 2018 Satellite Events, Heraklion, Crete, Greece, June 3–7, 2018, Revised Selected Papers 15* (Springer International Publishing), pp. 117–120.

Pandey, A. K., & Roy, S. S. (2024). Extractive question answering over ancient scriptures texts using generative AI and natural language processing techniques, *IEEE Access.* pp. 101197–101208.

Papavassileiou, K., Kosmopoulos, D. I., & Owens, G. (2023). A generative model for the Mycenaean Linear B script and its application in infilling text from ancient tablets, *ACM Journal on Computing and Cultural Heritage*, 16(3), pp. 1–25.

Raffel, C., Shazeer, N., Roberts, A., Lee, K., Narang, S., Matena, M., ... & Liu, P. J. (2020). Exploring the limits of transfer learning with a unified text-to-text transformer, *Journal of Machine Learning Research*, 21(140), pp. 1–67.

Reddy, G. (2023). The mechanistic basis of data dependence and abrupt learning in an in-context classification task, *Twelfth International Conference on Learning Representations*.

Reimers, N., & Gurevych, I. (2019). Sentence-bert: Sentence embeddings using siamese bert-networks, arXiv preprint arXiv:1908.10084.

Riemenschneider, F., & Frank, A. (2023). Exploring large language models for classical philology, arXiv preprint arXiv:2305.13698.

Ringe, D., Warnow, T., & Taylor, A. (2002). Indo-European and computational cladistics. *Transactions of the Philological Society*, 100(1), pp. 59–129.

Smidt, G. R., Lefever, E., & de Graef, K. (2024). At the crossroad of cuneiform and NLP: Challenges for fine-grained part-of-speech tagging. *Proceedings 2024 Joint International Conference on Computational Linguistics, Language Resources and Evaluation* (LREC-COLING 2024), pp. 1745–1755.

Snyder, B., Barzilay, R., & Knight, K. (2010). A statistical model for lost language decipherment. *Proceedings 48th Annual Meeting of the Association for Computational Linguistics*, pp. 1048–1057.

Sommerschield, T., Assael, Y., Pavlopoulos, J., Stefanak, V., Senior, A., Dyer, C., ... & de Freitas, N. (2023). Machine learning for ancient languages: A survey, *Computational Linguistics*, 49(3), pp. 703–747.

Ströbel, P. B., Clematide, S., Volk, M., & Hodel, T. (2022). Transformer-based HTR for historical documents, arXiv preprint arXiv:2203.11008.

Tangherlini, T. R., & Chen, R. (2024). Travels with BERT: Surfacing the intertextuality in Hans Christian Andersen's travel writing and fairy tales through the network lens of large language model-based topic modeling. *Orbis Litterarum*, pp. 1–44.

Uthirapathy, S. E., & Sandanam, D. (2023). Topic modelling and opinion analysis on climate change Twitter data using LDA and BERT Model, *Procedia Computer Science*, 218, pp. 908–917.

Vaswani, A., Shazeer, N., Parmar, N., Uszkoreit, J., Jones, L., Gomez, A. N., ... & Polosukhin, I. (2017). Attention is all you need, *Advances in Neural Information Processing Systems*, p. 30.

Volk, M., Fischer, D. P., Fischer, L., Scheurer, P., Ströbel, P., Sprugnoli, R., & Passarotti, M. (2024). LLM-based machine translation and summarization for Latin. *Third Workshop on Language Technologies for Historical and Ancient Languages – LT4HALA (LREC/COLING)*, Torino, 25 May 2024.

Wang, R., & Shi, Y. (2022, February). Research on application of article recommendation algorithm based on Word2Vec and Tfidf, *2022 IEEE International Conference on Electrical Engineering, Big Data and Algorithms (EEBDA)*, pp. 454–457.

Wei, J., Wang, X., Schuurmans, D., Bosma, M., Xia, F., Chi, E., ... & Zhou, D. (2022). Chain-of-thought prompting elicits reasoning in large language models, *Advances in Neural Information Processing Systems*, 35, pp. 24824–24837.

Williams, A., Nangia, N., & Bowman, S. R. (2017). A broad-coverage challenge corpus for sentence understanding through inference, arXiv preprint arXiv:1704.05426.

Yamshchikov, I. P., Tikhonov, A., Pantis, Y., Schubert, C., & Jost, J. (2022). BERT in Plutarch's Shadows, arXiv preprint arXiv:2211.05673.

Yang, Y., Theisen, R., Hodgkinson, L., Gonzalez, J. E., Ramchandran, K., Martin, C. H., & Mahoney, M. W. (2023). Test accuracy vs. generalization gap: Model selection in NLP without accessing training or testing data, *Proceedings 29th ACM SIGKDD Conference on Knowledge Discovery and Data Mining*, pp. 3011–3021.

Zuo, X., Zhang, S., & Xia, J. (2017). The enhancement of TextRank algorithm by using word2vec and its application on topic extraction, *Journal of Physics: Conference Series*, 887(1), p. 012028.

Part 3
Phylogenetic Methods

Chapter 4

Phylogenetic Trees, Reconstruction, and Validation Methods

4.1 Introduction

Phylogenetic approaches play a central role in understanding biologically evolving phenomena, like the evolution of plants, animals, or humans. The Tree of Life tries to represent the evolutionary relationships between all living things on Earth. It has three main branches representing the three domains of life: Bacteria, Archaea, and Eukarya. The concept of evolution also provides a powerful framework for understanding changes and adaptations in various domains outside of biology. Phylogenetic trees model the evolution of languages, documents, folklore, myths, cultural development, or patterns of inventions.

A phylogenetic tree is a branching graph that depicts the evolutionary relationships between different entities called taxa. Figure 4.1 shows an example. A taxon might be a species or a group of related species in biology. The taxa are represented as tree leaves, also called end nodes. Internal nodes are nodes that are not end nodes. In biology, each internal node represents the most recent common ancestor of all the taxa descending from that branching point. This ancestor is typically not directly observed in present-day data, but its existence is inferred based on the patterns of shared characters among its descendants. In phylogeny, one defines characters and states. For example, take the four taxa: eagle, cockatoo, lion, and horse. "Wing" is a character with state 0 in the case of lion and horse and 1 for cockatoo and eagle.

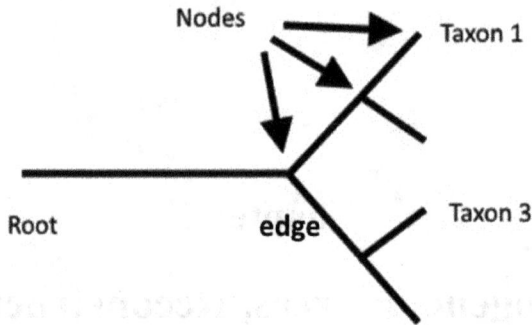

Fig. 4.1. Representation of a rooted phylogenetic tree with nodes corresponding to the end of one edge. The taxa are defined here on the end nodes. The root corresponds to the origin of the tree.

In a phylogenetic tree, the root represents the common ancestor of all the taxa on the tree. It is the point from which other nodes and branches stem. In a biological system, one can follow the branching patterns of the tree from the root to see how groups of organisms differentiate over time. In non-biological studies, the root may correspond to the common ancestor of some language family. The tree's root may be the origin or starting point of a cultural practice or tradition. In mythology or folklore, the root may be associated with a very ancient story that unfolds into many versions over time.

An unrooted tree is a phylogenetic tree without a specified root. It represents relationships between taxa without implying a particular direction of evolution. Unrooted trees are particularly useful when the root is uncertain or not well-supported by the available data.

Section 4.2 presents the ideal situation of a perfect phylogeny. The subsequent sections build on this important first section. The following sections explain how to reconstruct and validate a tree representing data that may not perfectly fit a tree topology and discuss various applications. The chapter concludes by exploring the potential of neural networks in phylogenetic studies.

4.2 Perfect Phylogenetic Trees

Perfect phylogenies provide a theoretical framework for understanding the methods and interpretations of phylogenetic analysis. This concept leads to straightforward validation algorithms, particularly well-suited to non-biological data. Understanding perfect phylogenies in some depth is important to explain why phylogenetic trees are robust against perturbations (lateral transfers in biology or borrowing in languages).

A perfect phylogenetic tree is extremely rare in practice. Nevertheless, the concept of perfect phylogeny is essential for understanding the diverse methods used to reconstruct and interpret phylogenetic trees in applications on imperfect data.

A perfect phylogeny is an idealized phylogenetic tree where each character state appears only once in the evolutionary history and is shared by all descendants of the ancestor in which it first appeared. A phylogeny defined by a set of characters is called a character-based phylogeny. A perfect rooted phylogeny is a phylogeny with each character state fulfilling the following two conditions:

- Each character state evolves only once.
- Any common character state is inherited from a common ancestor (i.e., defines a connected subtree).

These two conditions are equivalent to stating that each character is convex on the tree (Semple and Steel, 2002). For binary state characters (typically with states 0 and 1), convexity means that all taxa sharing a character state (e.g., all taxa with wings) are on a connected subtree, with the internal nodes having the same state as the taxa. It implies a single evolutionary origin for each state, meaning it evolved only once and was inherited by all its descendants. Figure 4.2 provides an illustrative

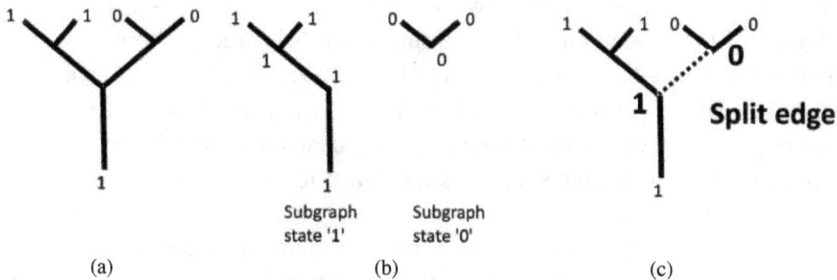

Fig. 4.2.   (a) The character with state zero and one is convex. After labeling the interior node, (b) one obtains two connected subtrees sharing no node. (c) The edge connecting the zero and one state node is a split edge. In a perfect phylogenetic tree, all characters are convex, and each edge is a split edge for some characters.

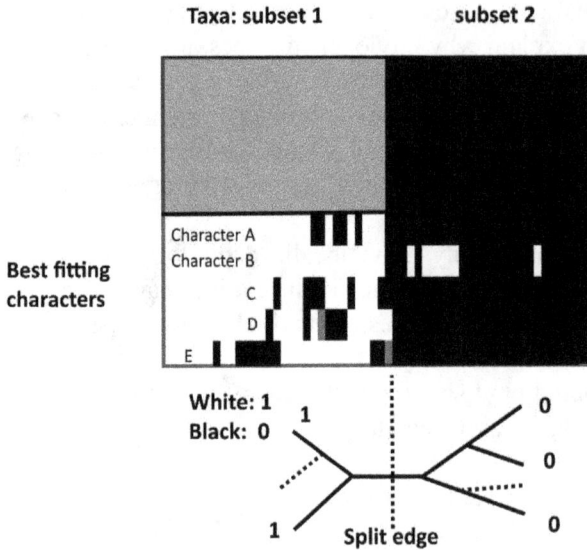

Fig. 4.3. A visualization of a bipartition in a phylogenetic tree (gray and black). The top rectangles represent the two subsets of taxa, while the bottom portion shows the five characters (A–E) that best fit this bipartition. The color-coding of these characters represents their state value. The dotted lines in the tree diagram highlight the split edge, which separates the two taxa subsets. The validation algorithm can identify such characters and assess their fitness to the bipartition, helping to evaluate the reliability of the tree topology.

example of a convex character on a tree. The edge connecting the zero and one state node is a split edge.

While perfect phylogenetic trees are rare in applications, the concept provides a valuable framework for analyzing real-world data. Figure 4.3 focuses on one edge defining a split in a tree (bottom). The top rectangles symbolically represent the two subsets as gray and black rectangles, ideally representing the bipartition of the taxa associated with the one edge. The five characters (A, . . ., E) best fitting the bipartition are represented by their states. As it is awkward to show many state values, the states are coded by color.

A simple validation algorithm for binary characters counts the deviation to a perfect bipartition of the character that best fits the bipartition. This simple validation algorithm allows us to identify the key characters that define the evolutionary relationships in the tree and assess how well those characters support the observed bipartitions. Characters fitting no bipartition associated with an edge can be seen as noise or a deviation from a tree topology.

Let us discuss two corollaries of the convexity requirement for binary state characters: the four-gamete rule and the consecutive-ones condition.

4.2.1 *The four-gamete rule and the consecutive-ones condition*

The four-gamete rule is central to understanding phylogenetic trees on binary characters. Consider four taxa having two binary state characters forming pairs, also called gametes: (0,0), (0,1), (1,0), and (1,1). Each taxon is characterized by one "gamete". The four-gamete rule states that no phylogenetic tree can perfectly represent the data if all possible combinations occur on two characters. Single (vertical) mutation events cannot explain their evolution, implying some lateral transfer or recurrent mutations (Semple and Steel, 2003).

The second corollary is the so-called consecutive-ones condition for binary state characters. To understand the condition, one must introduce the concept of circular order. Given a planar tree representation, the circular order is the order of the end nodes as one moves clockwise on a circular curve relating the end nodes. The consecutive-ones condition is important as it characterizes as well perfect trees and phylogenetic networks discussed by later in this chapter. A perfect tree is such that all one state of a character must be consecutive in a circular order of the taxa on the tree (see Thuillard, 2023a, 2023b, for a more formal definition). For instance, in the previous example, the taxa "eagle" and "cockatoo" that have wings should be consecutive. The consecutive-ones condition is illustrated in Fig. 4.4. Each of the three binary states fulfills the consecutive-ones condition and defines accordingly a split edge.

4.2.2 *Validation of character-based phylogenies*

The theoretical framework introduced in the previous section generalizes to multiple-state characters (say α, β, and χ). Consider the character states α and β and the following transformation:

$$C_{\alpha,\beta} = \begin{cases} 1 & \text{state } \alpha \\ 0 & \text{state } \beta \\ ? & \text{Otherwise} \end{cases} \qquad (4.1)$$

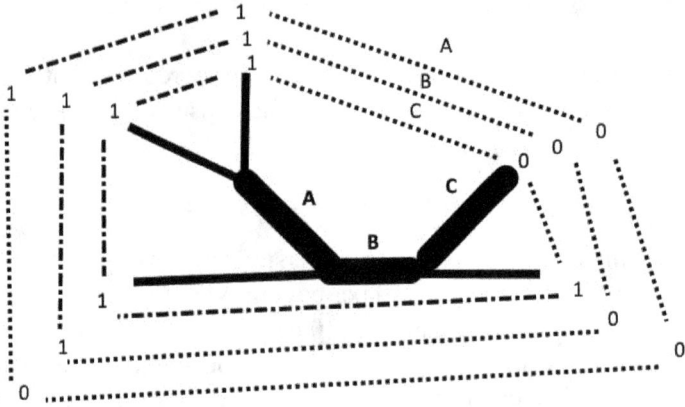

Fig. 4.4.   The three binary characters, A, B, and C, fulfill the consecutive-ones condition (the dot-dashed line connects the ones in a circular order of the taxa). Each character defines a split (labeled bold edge).

Stevens and Gusfield (2010) proved that in a perfect phylogeny, it is possible to define the value of each state "?" to either zero or one so that the phylogeny reduces to a phylogeny of binary states. Interestingly, the equation permits the integration of missing data by setting each missing state to the one or the zero-state value. Determining the value of each missing state "?" is complex. Invalidating the results is generally more simple. We suggest the following validation algorithm, which identifies deviations from a perfect phylogeny. Using Eq. (4.1), the algorithm searches for the character states that best fulfill each bipartition. This algorithm aims to identify deviations from a perfect phylogeny in binary and multistate character data.

Validation Algorithm:

1.   The algorithm searches for each character the states that fit best to a bipartition of a tree (associated with an edge) by minimizing the number D of character states deviating from a perfect bipartition.
2.   The minimum value D_i on all characters characterizes the deviation to a perfect phylogeny of a given bipartition of the ith character.

For bipartitions of very different sizes, the deviation can be weighted by the inverse of the cardinality of each subset in the bipartition ($1/n_1$, $1/n_2$)

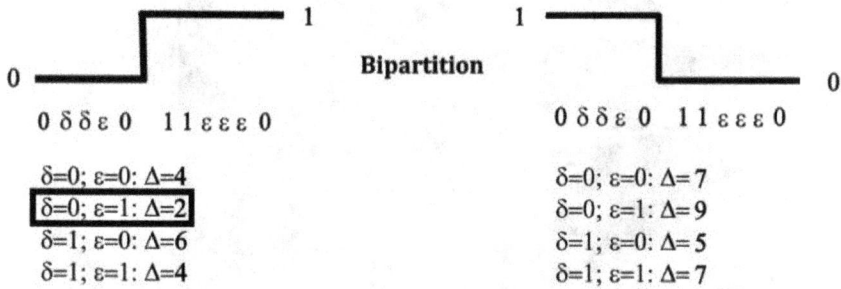

Bipartition

$0\ \delta\delta\varepsilon\ 0\quad 1\ 1\ \varepsilon\varepsilon\varepsilon\ 0$

$\delta=0;\ \varepsilon=0:\ \Delta=4$

$\boxed{\delta=0;\ \varepsilon=1:\ \Delta=2}$

$\delta=1;\ \varepsilon=0:\ \Delta=6$

$\delta=1;\ \varepsilon=1:\ \Delta=4$

$0\ \delta\delta\varepsilon\ 0\quad 1\ 1\ \varepsilon\varepsilon\varepsilon\ 0$

$\delta=0;\ \varepsilon=0:\ \Delta=7$

$\delta=0;\ \varepsilon=1:\ \Delta=9$

$\delta=1;\ \varepsilon=0:\ \Delta=5$

$\delta=1;\ \varepsilon=1:\ \Delta=7$

Fig. 4.5. Example of a 4-state character search for the best fit to a given bipartition. The minimum deviation to a perfect phylogeny has a value of two. One searches for the character and state for each tree bipartition, minimizing the error. The left and right bipartitions are the two possible instances to consider.

with $n_1 + n_2 = n$ with n the number of taxa or an entropy approach be used (see Eq. (4.6) in Section 4.3.4). Figure 4.5 shows an example with four states.

The algorithm's strength is that it only requires the input data and the tree with labels of the end nodes for validation. The deviation is zero if the phylogeny is perfect, but a zero deviation only guarantees a perfect phylogeny on binary state characters. For multistate characters, the validation algorithm furnishes a lower bound to the number of character states deviating from a perfect phylogeny. The algorithm may generate false positives for more than two states, but there are no false negatives for multistate characters. Its simplicity makes it easy to implement in phylogenetic trees and networks. We believe the approach is important as using global indices to characterize a result may not reflect how well some data fit the tree locally.

4.2.3 *Example of validation*

As an illustration of the validation algorithm, we have revisited a paper (Tehrani, 2013) on the diffusion of the famous tale of "Little Red Riding Hood" (ATU 333 for folktale specialists). The Grimm brothers' publication of a large collection of folktales triggered scholars' interest in folktales. Grimm and Grimm (1884) noted that many folktales were not particular to German languages but also found in very different traditions. The underlying assumption behind the discussed phylogenetic approach is

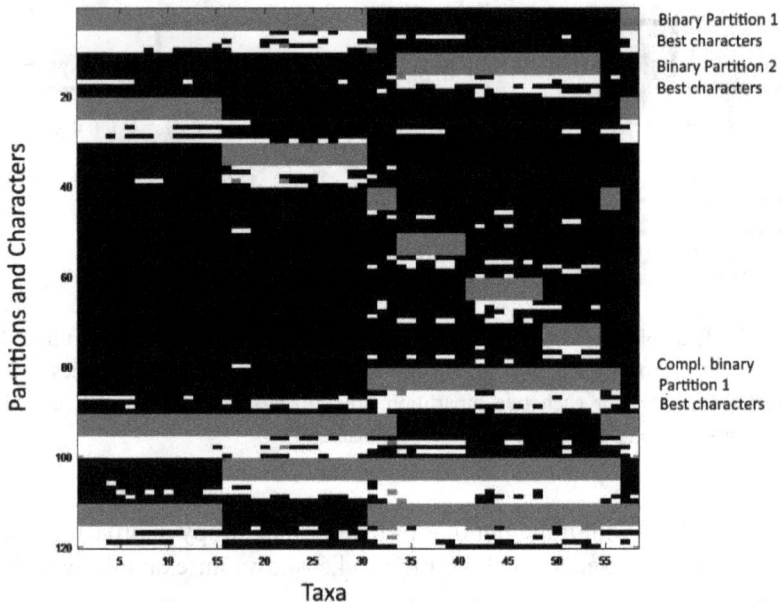

Fig. 4.6. Validation of the majority-rule consensus of the maximum parsimony trees of the "Little Riding Hood" article in Tehrani (2013). The black and gray bars correspond to the tree's main binary partitions associated with the main branches. The lines below the bars represent the five motifs with the lowest deviation, fitting the bipartition best (state 1: white; state 0: unknown: light gray). The top-left branch shows a clear branching. The remaining complementary binary partitions are not shown (high deviation values).

that if data fit a tree, then its different branches describe the 'historic' evolution of the tale. Figure 4.6 illustrates the application of the validation algorithm to the phylogenetic analysis of the "Little Red Riding Hood" folktale (The tree was generated using a majority-rule consensus of maximum parsimony trees; see Section 4.4 for details on these methods).

It reproduces the analysis of a single edge in Fig. 4.2 on all edges. The two subsets, defined by a bipartition, are gray and black segments corresponding to the taxa with identical states.

The lines below the bars represent the five motifs (characters) that best fit the bipartition, with white segments indicating the presence of a motif, black its absence, and gray indicating an unknown state.

The top-left branch shows a clear example of a bipartition where the characters strongly support the grouping of taxa. The right tree branches are supported but not as well as the left ones. The visualization provides a clear and intuitive way to assess the fit between the data and the tree, highlighting which characters are most important in defining the evolutionary relationships.

4.2.4 *On the robustness of the circular order of a tree against lateral transfer*

Phylogenetic trees are widely used outside of biology to describe the evolution of some cultural traits. Cultural elements may be transmitted within a population or through cultural interaction between traditions. The expression' lateral (or horizontal) transfer' describes the transmission of characters between individuals without a direct vertical inheritance from parents to their offspring. A lateral transfer from taxon A to taxon B corresponds to replacing some character state in taxon B with the ones of A. In a language tree, a lateral transfer may correspond to borrowing some words, like "weekend" in French.

In the previous section, we explained that a perfect phylogenic tree on multistate characters reduces through a transformation to binary state characters describing a phylogenetic tree. In other words, all transformed character pairs fulfill the four-gamete rule and the consecutive-ones conditions. A lateral transfer between consecutive taxa on a circular order may lead to the presence of the four gametes, disrupting a perfect tree. Conversely, the transfer of a character state preserves the consecutive-ones condition (Thuillard, 2008, 2009).

Figure 4.7 shows an example of a matrix that fulfills, after ordering, the circular consecutive-ones property. The matrix represents character states for different taxa. After ordering the taxa, all "one" states are

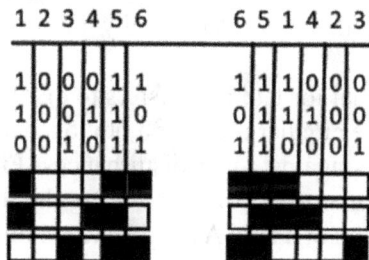

Fig. 4.7. After ordering the original matrix columns (left), each character (right) is so that all "one" states (one=black) are consecutive in a circular order (i.e., the first character is consecutive to the last one in a circular order).

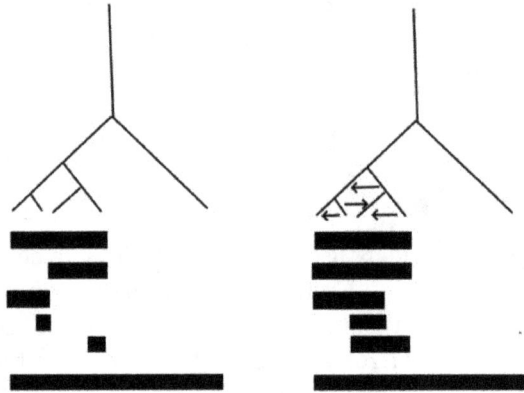

Fig. 4.8. Example of lateral transfers preserving the circular order of the tree. (Left) Without transfer; (Right) with lateral transfer.

consecutive in a circular order. The first character is consecutive to the last one in a circular order.

For binary characters, one readily understands (see Fig. 4.8 for an illustration) that lateral transfers between consecutive nodes in a planar tree representation do not disrupt the consecutive-ones property.

More generally, the circular order of a phylogenetic tree is robust against lateral transfer. Lateral transfers between consecutive taxa disrupt the tree structure but preserve the consecutive-ones condition in the transformed space using binary-state characters.

A set of ordered taxa satisfying the consecutive-ones condition can be perfectly represented as a phylogenetic network. The perfect network can be reconstructed using efficient algorithms (Bryant and Moulton, 2004) known as NeighborNet (see Section 4.3.3). No two edges cross in a phylogenetic network of this type, technically known as an outer planar network.

Despite the potential for lateral transfers to disrupt a perfect phylogenetic tree structure, the circular order of a phylogenetic tree demonstrates remarkable robustness against such disturbances. Figure 4.9 shows the robustness of the circular order of a tree using the degrees of liberty on the planar representation of a tree. A tree is like a mobile hanging from the ceiling. The branches can be moved around without changing the overall structure of the mobile. This flexibility in rearranging the branches represents the "degrees of liberty". All lateral transfers are after taxa reordering

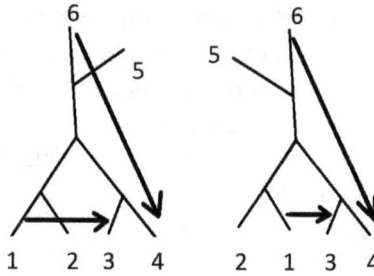

Fig. 4.9. By reordering the end nodes, there is a circular order so that all lateral transfers represented by an arrow are between end nodes adjacent on a circular order. The distance matrix associated with the perturbed tree fulfills the Kalmanson inequalities (Eq. (4.4)), and an outer planar network can exactly represent the distance matrix. Adapted from Thuillard (2008).

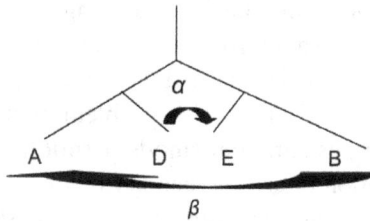

Fig. 4.10. A perfect order cannot be guaranteed if there is no order for all lateral transfers between adjacent taxa on the tree. Each arrow corresponds to a lateral transfer.

between consecutive nodes. An outer planar network perfectly describes the tree with the lateral transfer. The robustness against lateral transfers also has some limits.

A simulation study (Greenhill *et al.*, 2009) showed that the tree topology is generally not disrupted by up to 15% lateral transfer per taxon, concluding that tree topologies constructed with Bayesian phylogenetic methods are robust to realistic borrowing levels. Some warning should complete this statement. Figure 4.10 shows that the perfect tree topology is already disrupted with two lateral transfers, and there is no reason to believe that a Bayesian approach is fundamentally different here. In that example, there is no order in which all lateral transfers are between adjacent trees on the tree, and no outer planar network describes perfectly the tree with lateral transfers.

Concretely, lateral transfers are an important disrupting factor in the computation of deep branches or ancestor relationships. Nevertheless, they preserve neighboring relationships between taxa, provided the lateral transfers are between adjacent taxa in a circular order. In that case, an outer planar network perfectly describes the data. Section 4.3 explains how to construct the network!

4.3 Distance-based Phylogenies

For binary state characters and, therefore, for most applications outside biology, distance-based approaches furnish comparable results to the best approaches with the possibility of displaying the results as a tree or a phylogenetic network. Distance-based methods are generally faster than character-based methods (like maximum parsimony or maximum likelihood, ML), making them suitable for analyzing large datasets. The main weakness of distance-based approaches is that the phylogenetic information is summarized in a pairwise distance matrix, and therefore much phylogenetic information is lost. The problem is much less serious in binary state characters, as simple methods permit the visualization of the phylogenetic information after tree reconstruction (see Chapter 5).

In practical applications, data rarely fulfills a perfect phylogeny well. For instance, DNA bases (A, C, T, G) may regularly mutate in genetic applications. Frequent mutations can lead to large deviations from a perfect phylogeny. To address this, researchers often use a probabilistic approach (models of sequence evolution) or define a distance between taxa based on many characters (Jukes and Cantor, 1969). Distance-based approaches characterize the pairwise distance between any pair of two taxa. A phylogenetic tree is reconstructed from the resulting matrix. Neighbor-joining is the classical approach!

4.3.1 *Tree reconstruction with neighbor-joining*

Neighbor-joining (Saitou and Nei, 1997) is the standard algorithm to construct a tree from a distance matrix whose elements $d_{i,j}$ describe the distance between two taxa. The greedy algorithm connects two neighboring subtrees (or clusters of taxa) with an edge at each step.

Neighbor-Joining

Step 1: *Join neighboring clusters of taxa.*

Compute for each pair of taxa (i, j) the value $Q(i, j)$:

$$Q(i,j) = (n-2)\, d(i,j) - \sum_{i \neq k} d(i,k) - \sum_{j \neq k} d(j,k)$$

where n is the number of taxa (or clusters), and d is the pairwise distance matrix. Determine i_{max}, j_{max} minimizing Q.

Step 2: *Compute the distance between the new node i_{new} joining the two subtrees,*

$$d_{i_{max}, i_{new}} = \frac{1}{2} d_{i_{max}, j_{max}} + \frac{1}{2(n-2)} \left(\sum_{k=1}^{n} d_{i_{max}, k} - \sum_{k=1}^{n} d_{j_{max}, k} \right)$$

$$d_{j_{max}, i_{new}} = \frac{1}{2} d_{i_{max}, j_{max}} + \frac{1}{2(n-2)} \left(\sum_{k=1}^{n} d_{j_{max}, k} - \sum_{k=1}^{n} d_{i_{max}, k} \right)$$

Step 3: *Compute the distance between the new and other clusters.*
Use the reduction formula to replace the taxa i_{max}, j_{max} with the new taxon i_{new}:

$$d_{i_{new}, k} = \frac{1}{2} \left(d_{i_{max}, k} + d_{j_{max}, k} - d_{i_{max}, j_{new}} \right)$$

Different approaches have been proposed to actualize the distance matrix (Gascuel and Steel, 2009). If there is one perfect tree, the algorithm reconstructs it, while otherwise, it furnishes a good approximation.

4.3.2 *Tree validation*

The four-gamete rule cannot be applied to a distance-based tree to validate a phylogeny, as it relies on character states rather than distances between taxa. Instead, the four-point condition can be used to ensure that the

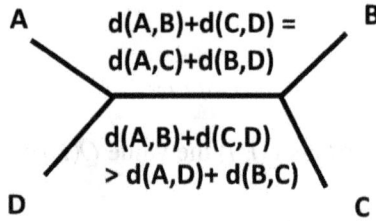

Fig. 4.11.   A tree defined by a distance is perfect if the four-point conditions hold on each quartet of taxa.

pairwise distances are compatible with an unrooted tree. The four-point condition states that for any four taxa A, B, C, and D, the following relationship should hold:

$$d(A, B) + d(C, D) = \max(d(A, C) + d(B, D), d(A, D) + d(B, C)). \quad (4.2)$$

Figure 4.11 illustrates the condition and furnishes a visual proof of Eq. (4.2). For binary state characters, Eq. (4.2) does not hold if the four gametes are present.

For distance-based phylogenies, we also need a method to verify that the distances between taxa are compatible with a tree structure, similar to how the consecutive-ones condition works for characters. The analysis is done on a transformed distance matrix. Given a reference taxon n, one computes the transformed distance matrix $Y_{i,j}^n$. Each matrix element corresponds to the distance from the reference taxon n to the path relating two taxa, i and j. The reader can verify in Fig. 4.12 that Eq. (4.3) gives the distance:

$$Y_{i,j}^n = \frac{1}{2}(d_{i,n} + d_{j,n} - d_{i,j}),$$

$$(4.3)$$

the transformed distance matrix describing a phylogenetic tree must fulfill the Kalmanson inequalities. For taxa indexed according to a circular order, the distance matrix describing a phylogenetic tree fulfills the so-called Kalmanson inequalities (Kalmanson, 1975):

$$Y_{i,j}^n \geq Y_{i,k}^n \geq 0 \quad i \leq j \leq k,$$

$$(4.4a)$$

$$Y_{k,j}^n \geq Y_{k,i}^n \geq 0 \quad i \leq j \leq k.$$

$$(4.4b)$$

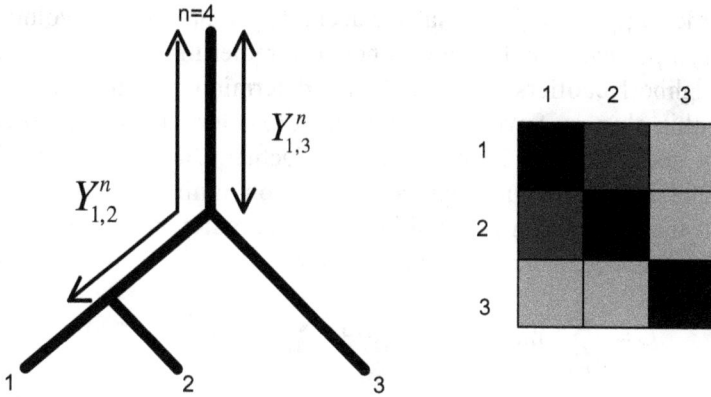

Fig. 4.12. The distance $Y_{i,j}^n$ gives the distance between a reference taxa n and the path $i-j$ on the tree. If the values of the distance matrix are coded in a grayscale, the level of gray decreases as one moves away from the diagonal. For more details, see Thuillard (2007).

The distance matrix $Y_{i,j}^n$ has the property that the distance diminishes away from the diagonal (Kalmanson, 1975). This property is visualized in Fig. 4.12. If different levels of gray represent the values of the distance matrix, then the colors are shading away from the diagonal. The relationships hold for data fitting perfectly a phylogenetic tree or network defined by distances.

The deviations to the Kalmanson inequalities permit the localization and visualization of the deviations to a perfect tree. A measure of the deviation to a perfect tree defined by a distance matrix is given by $max(Y_{i,k}^n - Y_{i,j}^n, 0)$ $i \le j \le k \le n$ which one calls the contradiction, and for which there are several variants (Thuillard, 2007, 2008). Let us note that for binary state characters, the Kalmanson inequalities are equivalent to the consecutive-ones condition.

4.3.3 *Distance-based phylogenetic networks*

NeighborNet, like the Neighbor-Joining algorithm for trees, reconstructs a phylogenetic network from a distance matrix. However, NeighborNet has a key advantage: it can reconstruct perfect networks and trees, making it a more general solution. Both algorithms iteratively join pairs of taxa or clusters with the Neighbor-Joining algorithm. NeighborNet includes an extra step to optimize the order of taxa within those joined clusters.

This ordering process is crucial for accurately representing evolutionary relationships when the data have a network rather than a tree structure.

NeighborNet offers some freedom in determining the optimal order of taxa within clusters. Several methods exist to determine the best ordering (Bryant and Moulton, 2004; Levy and Pachter, 2011; Thuillard, 2007, 2008, 2023a, 2003b). One approach focuses on minimizing the contradictions to the Kalmanson inequalities. The contradiction C can be quantified. For instance, one common measure is

$$C = \sum_{i<j<k<n} \max(Y_{i,k}^n - Y_{i,j}^n, 0) + \sum_{i<j<k<n} \max(Y_{k,i}^n - Y_{k,j}^n, 0), \qquad (4.5)$$

where i, j, and k represent taxa within the two joined clusters, and n is a taxon outside those clusters.

The algorithm searches for the order, minimizing the contradiction C at each iteration. The contradiction is zero for a perfect phylogeny.

Let us discuss the minimum contradiction approach in the context of binary states, which is the most common situation in a non-biological setting. For binary state characters, the reader can verify that the contradiction is zero if all the ordered character states fulfill the consecutive-ones conditions (The taxa are ordered so that the reference taxon n has the higher indices).

Figure 4.13 shows the different situations encountered in one ordering step. As Neighbor-Joining and NeighborNet have the same joining algorithms, the fulfillment of the consecutive-ones condition only depends on the ordering algorithm. At each step, the algorithm chooses an order minimizing the contradiction considering the two clusters and all reference taxa n. The algorithm furnishes a circular order for data fitting perfectly a phylogenetic network with no contradiction.

One of the main criticisms of distance approaches is that computing the distance matrix eliminates much information. This criticism does not hold for binary state characters, as the result can be analyzed separately for each character.

Technical note: Fig. 4.13 illustrates that the algorithm must choose an order where taxa with the same state are consecutive. However, if all characters within each cluster have the same state, the two clusters' taxa order does not affect the consecutive-ones condition due to symmetry. In other words, if at least one character in a cluster has two different states,

Ordering Step

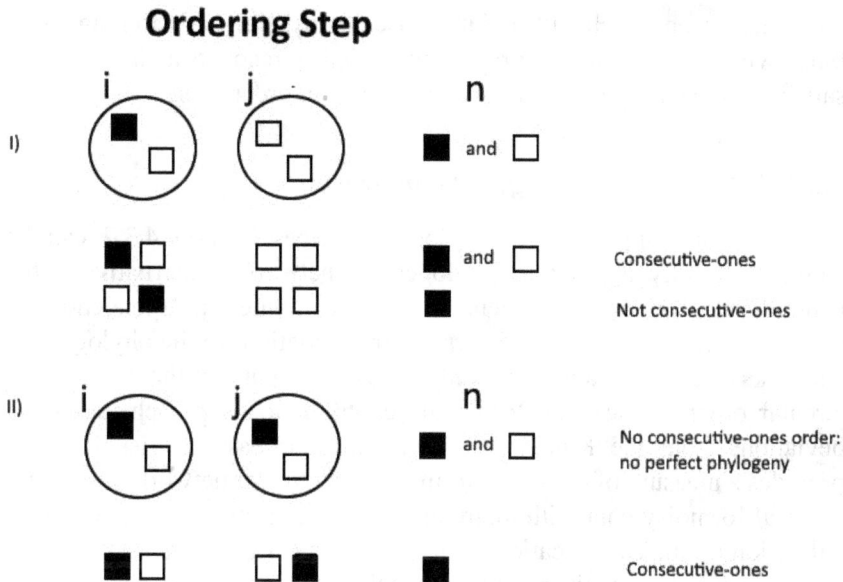

Fig. 4.13. Ordering of binary state characters fitting exactly a perfect phylogenetic network. The figure sketches one step of the ordering algorithm for non-trivial use cases. A square represents one or several binary state characters with identical states. The two joined clusters are i and j. The other clusters, marked as n, are all possible reference taxa. In case (i), the order of the taxa associated with node i is the first one, as the second order will lead to a deviation of the consecutive-ones condition. In case (ii), no perfect order exists if the subset of reference taxa contains both character states. The last line shows a perfect ordering if only one state is included. All other cases are either trivial or reduce to the above cases.

the algorithm must select the order that satisfies the consecutive-ones condition for all characters.

Bryant and Moulton (2004) showed that an order satisfying the consecutive-ones condition always exists for a perfect phylogenetic network. The minimum contradiction approach explicitly seeks to satisfy the consecutive-ones condition. This procedure guarantees that at each step, the two-joined clusters will satisfy one of the configurations in Fig. 4.13, for which a zero-contradiction ordering exists.

Therefore, when a contradiction-free order exists (as is the case for perfect phylogenetic networks with binary data), the minimum contradiction approach will find it. Consequently, Neighbor-Joining, combined with the minimum contradiction approach, recovers a circular order satisfying the circular-ones condition if such an order exists.

4.3.4 *Validation of phylogenetic networks*

The validation approach for phylogenetic trees Section 4.3.2 can be straightforwardly applied to phylogenetic networks. Alternatively, the contradiction on each character can be computed independently of the phylogenetic structure to determine the deviations to the phylogenetic networks of each character. The only necessary inputs are the data and the circular order of the taxa. While the contradiction approach, based on deviations from the Kalmanson inequalities, is easy to visualize and provides a measure of deviation from a perfect tree or network, it may not be ideal for noisy data with many missing data or errors. An example of validation on the classification of myths is furnished in Chapter 5.

In such cases, an alternative approach based on entropy validates the resulting tree or network topology. Given a bipartition in the tree, we can compute the entropy of each taxa subset defined by the split. Entropy measures the disorder within a subset:

$$H(Y) = -p * \log_2(p) - (1 - p) * \log_2(1 - p), \qquad (4.6)$$

where p is the proportion of one state. The entropy of a subset with all equal states is zero. To characterize a bipartition, we calculate the entropy for both subsets created by the split and use the highest value between the two or some average. A value below some threshold may serve as a validation criterion. Let us note that entropy also characterizes the correlation between a trait and a phylogeny (Borges *et al.*, 2019) in Bayesian phylogenies (Ribeiro *et al.*, 2023).

A specialist will find validation methods at the character levels extremely useful. This approach can be applied to phylogenetic trees and networks. Large deviations from the consecutive-ones condition or large entropy values indicate much uncertainty in ancestor state reconstruction.

4.4 Maximum Parsimony

The maximum parsimony approach is conceptually very important to understand phylogenetic trees. William of Ockham (or Occam) was a

Franciscan friar born in England in 1285. He is best known for his "razor", a principle of parsimony that states that the explanation with the fewest assumptions should be selected among competing hypotheses. In other words, we should not use more assumptions than necessary to explain a phenomenon. The Occam principle is the basis of the maximum parsimony approach. Maximum parsimony is a character-based approach that constructs an evolutionary tree from characters known on a set of taxa. The best tree displays the smallest number of state changes explaining the observed character states. The Fitch (1970) and Sankoff (1985) algorithms are the most popular for finding the minimum number of state changes necessary to fit the data on a phylogenetic tree. The Fitch algorithm is simpler than the Sankoff one, and we limit the discussion to that one. The Fitch algorithm is a heuristic that suggests the most parsimonious state for each internal node of the tree. The algorithm combines a bottom-up and top-down approach, the bottom being the end node.

4.4.1 *Fitch algorithm*

Given a tree topology, the algorithm assigns a character state to each character at every internal node.

Stage 1: From the end nodes to the root
The algorithm lists the compatible states for each internal parent node. The compatible states are the union of the two parent states or their intersection if the two child nodes have some common states.

Stage 2: The algorithm chooses the character state from the root to the end nodes, minimizing the number of state changes.
Output: Number of character state changes
Figure 4.14 illustrates this stage with a simple example. A tree's parsimony score is the number of character state changes required to explain the observed data. The computationally intensive part involves searching for the tree topology with the lowest score.

Some character states might be unknown. In that case, one may try inferring the state from neighboring states or some interpolation scheme or eliminate the character. Another possibility is to attribute a missing state "?" to the character and to use the Fitch algorithm without modification. While methods for computing the parsimony of a tree are simple and computationally efficient, the number of trees grows exponentially with

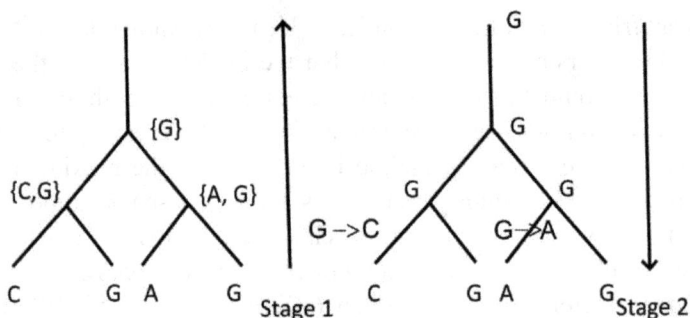

Fig. 4.14.   Illustration of the Fitch algorithm. (Left) The algorithm from the end nodes to the root; (Right) the second stage from the root to the bottom.

$n!$ ($n! = 1 * 2* \ldots * n$). The exhaustive search for the tree maximizing the parsimony becomes rapidly practically impossible, and heuristics are used instead. Let us recall that in data processing, a heuristic is designed to find a good, approximate solution to a problem when getting the best solution is computationally too expensive. Heuristics are generally based on common sense or experience and can effectively find good solutions in a reasonable amount of time.

4.4.2 *Tree validation*

The consistency and retention indices characterize how well the data fits the maximum parsimony trees. Following Farris (1989), the consistency index (CI) is an index on a single character but easily generalizes to all characters. The CI is the ratio between the smallest number of state transitions M for the character on any tree and the number of state transitions S in the most parsimonious tree.

$$CI = M/S. \tag{4.7}$$

The value S can be partitioned into observed vertical state transitions and homoplasies. Homoplasy refers to a state shared between two or more species that did not originate from their most recent common ancestor. The number of homoplasies H is given by

$$H = S - M. \tag{4.8}$$

A perfect phylogeny is homoplasy-free.

The retention index is defined as

$$RI = (G - S)/(G - M), \tag{4.9}$$

where G is the maximum number of homoplasies on any tree given the characters and states (Farris, 1989). The largest number of homoplasies during the most parsimonious tree search may estimate the value G. One may alternatively use an algorithm searching for maximum homoplasy instead of maximum parsimony. (For binary characters, if $G = M$, then the quotient is set to zero.) Figure 4.15 illustrates the approach.

The value of the retention index is between zero and one. An RI close to one indicates that the tree fits the data well. The retention index is a single index qualifying the "goodness" of a tree. One must be aware that

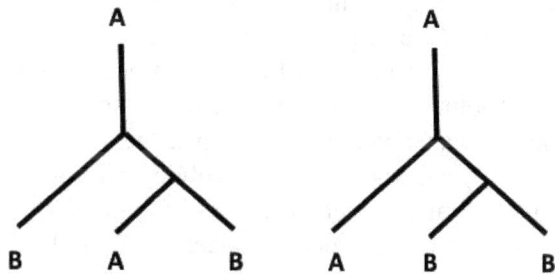

	Left tree	Right tree
M: Minimum number of state transitions on any tree	M=1	M=1
S: Number of state transitions on most parsimonuous tree	S=2	S=1
G: Maximum number of state transitions on any tree	G=2	G=2
CI: Consistency index CI= M/S	CI=0.5	CI=1
RI: Retention index RI= (G-S)/(G-M)	RI=0	RI=1

Fig. 4.15. The examples illustrate the computation of two trees' consistency and retention factors. (Left) For the character with state A or B, the tree has the most state transitions on any tree $G = S$; (Right) For the same character, the tree has no homoplasy ($S = M$ and $H = S–M = 0$).

the value may be deceptive in several manners. A tree may have a large retention index on the tree, though a subtree may have a low retention index. Therefore, we introduced other validation methods in Section 4.2 that allow for a finer analysis.

4.4.3   *Searching for the ancestral states*

Rooted trees are sometimes used to infer the root's original state. Considering the above discussion, this task may appear simple using the maximum parsimony approach. The tree minimizing the number of state changes is constructed by assessing the state of each internal node and the root. The character states at the root are, therefore, determined. This task is, in reality, quite tricky and prone to errors, some of them being difficult to find out. The most obvious problem is when the root has a state that does not belong to the set of states in the end nodes. In that case, the information on the origin state is lost. The Fitch algorithm will nevertheless suggest a (wrong) ancestral state among the set of end node states.

Another issue is when the root taxon is unknown, and one relies on choosing a taxon or an outgroup as the root. In that case, midpoint rooting is appropriate for characters evolving at a constant rate, for instance, when a molecular clock characterizes the rate of state changes. In its simplest version, midpoint rooting places the root at the midpoint of the longest branch in the tree based on the assumption that the two taxa at the tips of the longest branch are the most divergent and, therefore, the most likely to share a common ancestor at the root. Figure 4.16 illustrates midpoint rooting. The method does not work with different evolution rates on the branches.

The search for ancestral states is generally prone to errors if the number of interior nodes is much higher than the number of known end nodes or if there are missing taxa. Missing taxa may change the root states (see Fig. 4.17).

The most serious problem is when lateral transfers alter a phylogenetic tree, leading to a phylogenetic network structure. Lateral transfers may erase the past evolution on large tree parts, making ancestor state determination almost impossible. Searching for ancestor states is a legitimate inquiry, but the results must be carefully analyzed and should not rely solely on a phylogenetic analysis.

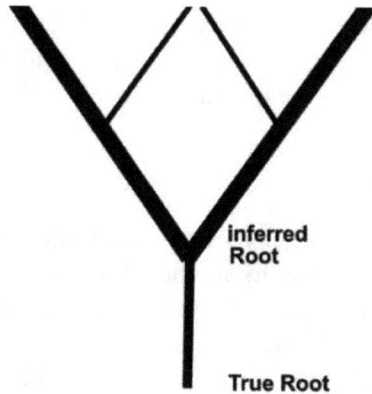

Fig. 4.16. Rooting a tree using the midpoint approach. In that example, the inferred root corresponds to the last common ancestor if one assumes the true root is at the position marked on the tree.

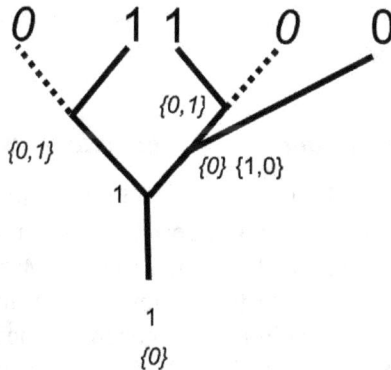

Fig. 4.17. The root state is zero on the tree (thick line) on the three taxa. Adding two unknown taxa may dramatically change the root state (Italic numbers and dotted line).

4.5 Maximum Likelihood (ML) and Models of Evolution

Due to its simplicity, maximum parsimony is a good option for very large problems. The approach is usually assumed to work well for rare state changes (Fischer, 2023). However, maximum parsimony does not consider the relative likelihood of character state transition; therefore, a ML or a Bayesian approach is generally preferred. These approaches model the transition rate between states, implicitly requiring some molecular clock. Let us now discuss the evolution models and the ML.

4.5.1 *Transition rates*

The ML approach to phylogeny uses a stochastic model to estimate the transition rate from one state to another. The approach assumes that the transition probability only depends on the present state. One speaks technically of a Markov chain describing a sequence of possible events where the probability of each event depends only on the state attained in the previous event. In that model, a matrix Q describes the rate of transitions between states. The element of the matrix Q corresponding to the ith row and the jth column furnishes the rate at which a state i transforms into a state j. The element $Q_{1,0}$ gives the transition rate from state 1 to state 0. One understands that a symmetric model with $Q_{1,0} = Q_{0,1}$ deviates from the assumptions of a perfect phylogeny as a state may appear several times during evolution.

4.5.2 *Maximum likelihood in phylogenetic trees*

The likelihood measures how well a given model with specific parameter values explains the observed data. It corresponds to the probability of the data given the parameters, or likelihood L: P(*Data|Model*). The ML principle states that the model with the parameters that maximize the likelihood function is the most likely to have generated the observed data. The likelihood of a phylogenetic tree is the probability of observing the given sequence data under that tree and a specific evolutionary model, including a molecular clock and the Q matrices for the different transition rates. For simplicity, let us write all parameters as $\vartheta = [\vartheta_1, \ldots, \vartheta_k]$. The ML parameters correspond to the parameters maximizing the likelihood function for the given data.

$$L(x, \vartheta). \tag{4.10}$$

For independent and identically distributed random variables, the likelihood is the product of function:

$$L(x, \vartheta) = \prod_{i=1}^{k} L(x, \vartheta_k). \tag{4.11}$$

In practice, the ML values can be very small, and using the natural logarithm of the likelihood is convenient. Since the logarithmic function monotonously increases, taking the logarithm does not modify the ML. For a differentiable function $L(x, \theta)$, the ML satisfies the condition:

$$\frac{\partial L}{\partial \theta_i} = 0. \tag{4.12}$$

Finding the maximum analytically is rarely possible, and numerical optimization using a specialized program or general software such as Matlab or R is often necessary.

There are two types of optimizations: the one on the tree topology and the one on continuous parameters. Each tree topology furnishes a likelihood, and the optimization searches on different trees by using some procedure to modify the tree, most often through the cut and paste of a branch. For continuous parameters like the length of a branch, the software programs implement numerical methods, such as the expectation–maximization (EM) algorithm in Section 7.4.2.2 (Ng *et al.*, 2012), to find the parameters that give the ML.

There are several essential aspects to consider when using a ML approach. While the algorithm is highly efficient, it can be computationally demanding. The number of candidate trees increases exponentially with taxa, limiting the number of taxa that can be analyzed. Additionally, the number of parameters can be a limiting factor. For more complex models, finding the values of the parameters with ML values can be computationally expensive. It is crucial to remember that the results may be misleading if the data do not fit a tree or the evolution model is incorrect.

Several software programs for ML computation exist. Some are freeware, some give a free university license, and others are commercial. Phylogenetic analysis using parsimony (PAUP) RAxML can efficiently infer phylogenetic trees from various data types, including binary characters, on large datasets (Kozlov *et al.*, 2019). It can estimate branch lengths and perform ancestral state reconstruction. IQtree is another much-used program.

4.5.3 *Molecular clock*

Dating major evolutionary events is a standard task in phylogeny. The basic idea behind molecular clocks is that the evolution rate of DNA, proteins, or, more generally, some characters is relatively constant over time. The branch length linearly relates to the time between the two nodes in the constant molecular clock model. This model is too simple and generally replaced by a distribution allowing clock variations. Typically, a clock value is drawn from a distribution and optimized to fit the data best. Applications outside of biology are also striving for molecular clocks. The transition rate between cognates is not well-known in linguistics and may vary significantly.

4.6 Tree Optimization

Phylogenetic methods, except distance-based approaches, rely on an algorithm to compute how well a tree fits the data. Computing all possible trees and finding the best solution is not computationally feasible for more than a few taxa. For that reason, methods have been developed to generate efficient new trees using a search heuristic. A heuristic search is an algorithm that decides how to search next using a strategy adapted to the problem. Heuristics are typically used in "hard" mathematical problems, like finding the tree associated with the maximum parsimony solution. PAUP stands for the acronym of one widespread phylogenetic software. The interface offers several possibilities for searching the tree topology, maximizing the parsimony. A heuristic implements proven strategies for some problems but does not guarantee finding the best solution. The tree re-arrangement strategies presented in this section apply to either a maximum parsimony, a ML, or a Bayesian approach. In order not to overload the text, the presentation is limited to the maximum parsimony. PAUP offers several heuristics:

- Nearest neighbor interchange (NNI): The algorithm exchanges pairs of adjacent branches in the tree.
- Subtree pruning and regrafting (SPR).
- Tree bisection and reconnection (TBR).

4.6.1 *Nearest-neighbor interchange*
NNI typically considers three subtrees at each step and swaps their connectivity. Figure 4.18 illustrates the process. The algorithm looks at the connectedness of the four subtrees represented by a quartet, defined as a set of four nodes with their associated subbranches. As shown at the

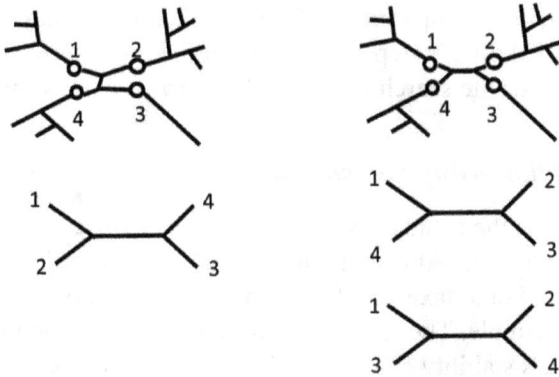

Fig. 4.18. Illustration of the NNI. The two top figures show an example of a transformation in the quartet. The three different alternatives are sketched below.

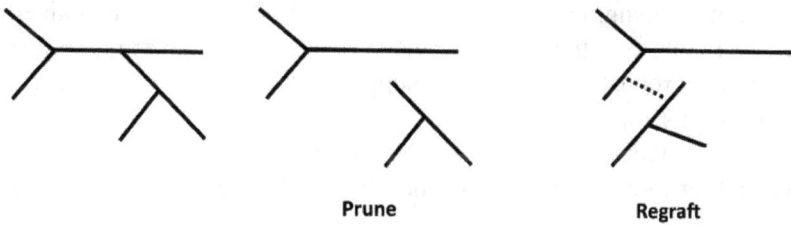

Prune   **Regraft**

Fig. 4.19. Subtree pruning and regrafting. An edge is removed, resulting in two subgraphs reconnected by an edge at some point.

bottom, three different ways of connecting the four subtrees exist. Each NNI operation generates two new trees.

4.6.2 *Subtree pruning and regrafting and tree bisection and reconnection*

Figure 4.19 illustrates the process of SPR. The tree is cut (pruned) into two subtrees at each iteration step by disconnecting an edge from its attached node. The disconnected edge is connected to an edge to form a new tree. The algorithm is more complex because the number of possibilities at each iteration step is larger than in the NNI. The TBR algorithm has one more degree of liberty as the edge connecting the two subbranches in the whole tree is removed, and the subtrees are reconnected with a new edge. Typically, the search algorithm accepts a new tree topology if it raises the parsimony. In some options, the algorithm may accept, with some low probability, a step that increases the parsimony. This strategy effectively avoids the search to get stacked in a local maximum.

4.6.3 *Validation using bootstrapping*

Bootstrapping is the standard verification method. The approach is computationally intensive. At each iteration, taxa are sampled randomly from the set of taxa. Some taxa will have more than one copy. Others will not belong to the sample. The resampling may also be done on the characters to test the tree's stability against small changes in characters. Each bootstrapping iteration results in a tree (Fig. 4.20). The support for each branch of the original tree corresponds to the number of trees with the presence of the branch. Each branch defines a bipartition of the set of taxa. Consider a branch of the original tree and the associated bipartition of the taxa. If, after bootstrapping, in most trees, some subset of taxa are all contained in the same subset as the original bipartition, then the corresponding split edge is very robust. The bootstrap support values indicate the percentage of bootstrap samples with the same branch.

An alternative is Jackknife, with resampling by removing one (or several branches) from the original set. Jackknife resamples the data

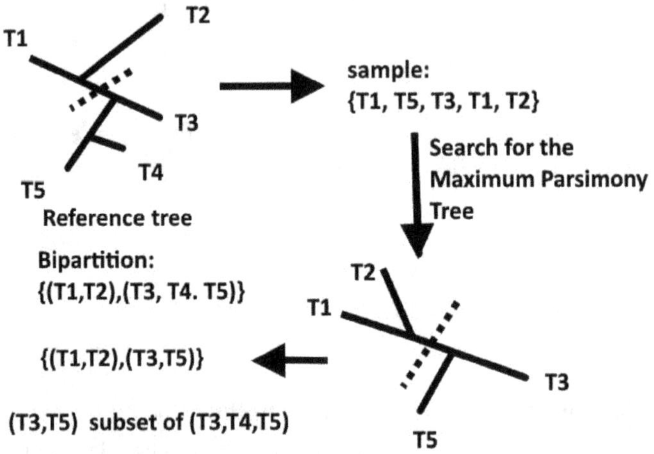

Fig. 4.20. The bootstrapping algorithm randomly samples taxa and searches for the best tree. For each edge, the algorithm compares the bipartition of the reference tree to the new best tree. The bipartition is robust in that sketch, though one branch (T4) was removed.

without replacement, meaning each taxon is selected at most once. The sample is smaller than the original dataset. The algorithm searches for a Maximum Parsimony tree for each sample. The support value for each branch is the percentage of samples in which the branch is present. Bootstrap is generally the method of choice.

4.7 Specialized Transition Models

While the Markov approach is at the core of almost all evolutionary models, the form of the matrix Q depends on the field of application (genetics, linguistics, paleontology, phylogeography). The next sections introduce the most important models.

4.7.1 *Historical linguistic*

The Pseudo-Dollo covarion model is typically used to model the evolution rates of different cognates in a language. For simplicity, we present the covarion model for binary state characters. The results are easily adapted to multistate characters. Step by step, let us introduce the Pseudo-Dollo covarion model.

For binary characters, the continuous time Markov chain (CTMC) states assume that a character has two possible states: absent or present. The rate matrix is symmetric and contains a single parameter, and all transition rates are constant:

$$Q = \begin{pmatrix} -\gamma & \gamma \\ \gamma & -\gamma \end{pmatrix}.$$

$$(4.13)$$

The symmetric matrix has a single parameter. Let us note that the sum of the values on each line and column is zero. (The sum of the probabilities in a Markov process equals one. The rate is obtained by differentiating the transition matrix; therefore, one expects the sum to be zero.) A simple diagram represents the possible transitions with binary state characters. Figure 4.21 shows two transitions from state 1 to state 0 and from state 0 to state 1. In the event of no transition, an arrow starting and ending at the same state represents the event of no transition.

The covarion model diverges from Eq. (4.13) as it postulates two modes of evolution: fast and slow (Fig. 4.22). The transition rate matrix

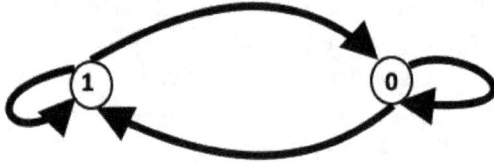

Fig. 4.21. Model describing the reversible transition between two binary states. Such a model is used to visualize transitions described by a Markov process in which the next state only depends on the present state.

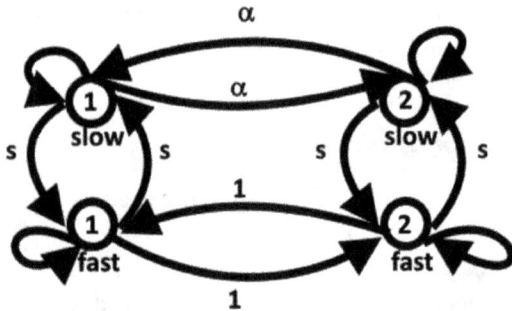

Fig. 4.22. The covarion model describes the transition between binary states. Over time, the characters switch back and forth between a fast and a slow mode.

Q is a 4 × 4 matrix for binary state characters. The fast switching rate between the two states is set to a fixed value of 1 (as we are here only interested in the tree's shape, setting one parameter to one is justified and consequently decreases the number of parameters by one). The slow transition rate equals α. The factor s describes the switching rate between the slow and fast modes. When the mode switches, one assumes that the state character stays identical and, therefore, the zeros in Eq. (4.14).

The model is used in molecular genetics. The covarion matrix models irregularities in the rate of language evolution. The covariation transition rate is given by

$$Q = \begin{pmatrix} -1-s & 1 & s & 0 \\ 1 & -1-s & 0 & s \\ s & 0 & -\alpha-s & \alpha \\ 0 & s & \alpha & -\alpha-s \end{pmatrix}. \tag{4.14}$$

The first two lines correspond to the fast mode, and the two last to the slow mode with a transition factor ($\alpha \ll 1$).

In many studies, the phylogenetic analysis of languages relies on a list of cognates. There are widespread words with a common origin, like foot in English and Fuss in German. Cognates are typically related etymologically. They descend from a common ancestor word in a proto-language. Cognates appear once; if they disappear during evolution, one assumes they are forever gone. Time-reversible transition matrices are inappropriate in that application. To remedy it, the pseudo-Dollo model has three states: initial, present, and removed. The transition matrix is of the form:

$$Q = \begin{pmatrix} -\lambda & \lambda & 0 \\ 0 & -\mu & \mu \\ 0 & 0 & 0 \end{pmatrix}. \tag{4.15}$$

Figure 4.23 shows the corresponding transition diagram:

The factor λ corresponds to the birth rate (generally set to one), μ the death rate, and the character has state "removed". The pseudo-Dollo model is about the evolution of states along the different lineages. The two models share a similar structure but apply to different aspects of an evolutionary tree. The pseudo-Dollo covarion model (Bouckaert and Robbeets, 2017) combines the covarion model with a slow transition and a fast transition rate with a birth and death model (Fig. 4.24). The pseudo-Dollo covarion transition rate is given by

$$Q = \begin{pmatrix} -1-s & 1 & s & 0 & 0 \\ 0 & -\mu-s & 0 & s & \mu \\ s & 0 & -\alpha-s & \alpha & 0 \\ 0 & s & \alpha & -\alpha-s & 0 \\ 0 & \mu & 0 & 0 & -\mu \end{pmatrix}. \tag{4.16}$$

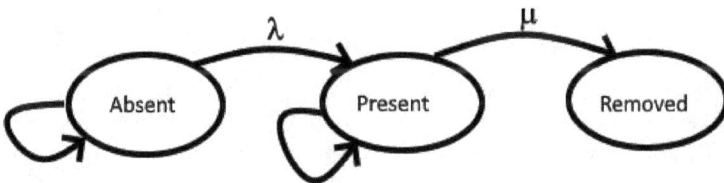

Fig. 4.23. The pseudo-Dollo model may describe a character state's apparition and disappearance.

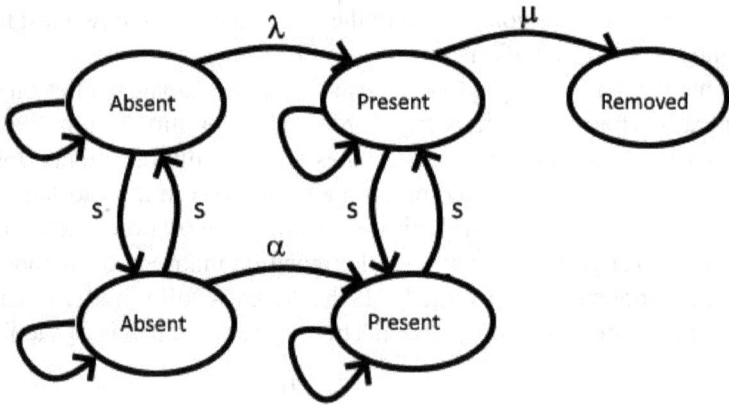

Fig. 4.24.   The pseudo-Dollo covarion model combines the covarion model with a slow transition and a fast transition rate with a birth and death model.

The first two columns correspond to the fast mode, and the two next ones to the slow mode with a transition factor ($\alpha \ll 1$). The last column describes the transition to the removed state. The matrix is not time-reversible as the matrix is not symmetric.

4.7.2 *Fossilized birth-death model*

The fossilized birth-death (FBD) model allows the integration of extant and extinct taxa fossils into phylogenies. Since fossils are typically not associated with molecular sequences, the FBD process requires combining data from various sources to estimate evolutionary parameters and reconstruct past events. In total approaches, the model integrates fossils' dating and morphology (Ronquist *et al.*, 2012). The fossilized birth and death model contains two supplementary parameters compared to the birth and death model (see Complement Section 4.11). The following four parameters describe the tree distributions:

Birth rate (λ): The rate at which new species arise from existing lineages.
Extinction rate (μ): The rate at which species go extinct.
Fossilization rate (ψ): The rate at which organisms are fossilized and preserved in the fossil record.
Sampling rate (ρ): The rate at which paleontologists discover and sample fossils.

The model's last parameter, ρ, describes the sampling probability of extant species. Stadler (2010) obtains an analytic formula for the conditional probability density of a tree $\mathcal{T}$: $f[\mathcal{T}|\lambda, \mu, \psi, \rho, x1]$. This expression summarizes the probability of the topology, internal node ages, and fossil attachment times in $\mathcal{T}$. The probability is conditional on the parameters $(\lambda, \mu, \psi, \rho)$ and the age of the root node $(x1)$. Fossils can also be integrated into the tree model, setting constraints on the topology (Heath *et al.*, 2014; Zhang *et al.*, 2016; Barido-Sottani *et al.*, 2023).

The Fossilized Birth and Death model can be extended to morphological changes like the absence or presence of an organ or states describing very different morphologies. The Mk model is often used to model the evolution of morphological characters. The Mk model is, in spirit, very similar to the Jukes-Cantor model, as it assumes equal transition probabilities between states. After binning the continuous state values, a transition rate matrix describes possible transitions. The Mk matrix (Lewis, 2001) is a straightforward generalization of the Juke-Cantor matrix to an arbitrary number of states. For k states, one has

$$Q = \begin{pmatrix} -(k-1)\mu & \mu & \dots & \mu & \mu \\ \mu & & \dots & & \mu \\ \dots & \dots & \dots & \dots & \dots \\ \mu & \mu & \mu & \mu \\ \mu & \mu & \mu & -(k-1)\mu \end{pmatrix}. \tag{4.17}$$

If large fluctuations of morphological values are not expected, the following transition matrix can be applied:

$$Q = \begin{pmatrix} -\mu & \mu & 0 & 0 \\ \mu & -\mu & \mu & 0 \\ 0 & \mu & -\mu & \mu \\ 0 & 0 & \mu & -\mu \end{pmatrix}. \tag{4.18}$$

The above matrices are easily integrated into a Bayesian scheme to compute the likelihood of the different states. In a Markov framework, one multiplies the likelihood and prior probabilities for the various processes, resulting in very small numbers. For that reason, the logarithm of the probabilities is generally computed to replace the multiplications by sums, which are handled better by a computer.

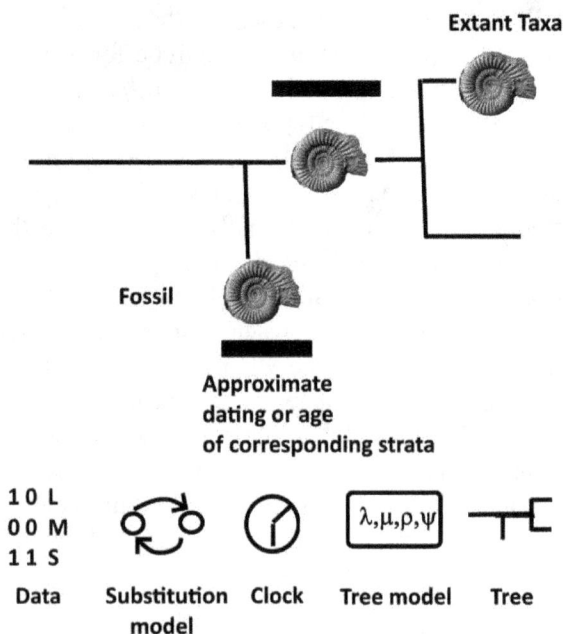

Fig. 4.25. Total-evidence model using the Fossilized Birth and Death model. The phylogeny is modeled using a substitution model, a molecular clock, and a fossilized birth and death process. Fossils are added to the model, and morphology evolution is integrated through a Mk model.

Figure 4.25 summarizes the process. The input includes both molecular and morphological data, the likelihood of which can be computed using substitution models with their corresponding transition rates Q, a clock, and a tree model.

4.8 Some Applications

Phylogenetic trees are important in classifying living organisms and, among other things, studying the domestication of plants and animals. They may also represent relations between manuscripts, languages, motifs, comparative mythology, archaeology, or cultural studies.

4.8.1 *Stemmatology and historical linguistic*

Stemmatology studies how texts evolve and change, particularly in manuscript traditions. Phylogenetics is one of the methods used in that field, with most studies using a maximum parsimony approach for similar versions and a distance approach for analyzing very different versions, for instance, to classify the different versions of the Canterbury Tales (Barbrook *et al.*, 1998). Word embedding, or sentence embedding (see Chapter 3), is sometimes used to preprocess data before reconstructing the genealogy of different versions. Sanskrit texts (Kanojia *et al.*, 2019) were classified using a distance matrix obtained by pairwise multiplying vectors characterizing documents. Another approach applies the TF-IDF method to create a distance matrix. This method calculates the importance of words in a document relative to a larger corpus (Marmerola *et al.*, 2016). The resulting distance matrix becomes the input to a tree reconstruction method. Recent studies have explored phylogenetic networks and extended studies on phylogenetic trees using ML and Bayesian methods. Hyytiäinen (2022) compared the different versions of a chapter of the Acts of Apostles. The study finds that the different versions are too complex to be described by a single phylogenetic tree and that a phylogenetic network is best suited to the task.

A similar approach, using phylogenetic networks, is also possible on music pieces. Windram *et al.* (2022) analyzed two preludes from J.S. Bach's Well-Tempered Clavier II. It showed that Bach did not modify one prelude much but considerably reworked the second one. McCollum and Turnbull (2024) apply an ML and a Bayesian approach to versions of the Epistle to the Ephesians using the Mk-model, in which all changes are equiprobable. The study shows proximity between some versions, but the closer to the root, the larger the uncertainty on the relationships between texts.

The reader will find a review of the field and further applications in Howe and Windram (2011, 2023).

The evolution of language families is often described with phylogenetic trees or networks (Ringe *et al.*, 2002; Barbançon *et al.*, 2013; Nakhleh *et al.*, 2005). Chapter 6 presents the latest developments, and we refer to Pellard *et al.* (2024) and Heggarty *et al.* (2023, supplementary material) for a very good presentation of the historical developments in the last 20 years, with its highs and lows.

Let us mention an interesting observation. Canby *et al.* (2024) found that explicitly accounting for lexical polymorphism produced more accurate results than even the most sophisticated tree reconstruction methods, such as the Gray-Atkinson Bayesian approach (Gray and Atkinson, 2003). The author defines lexical polymorphism as a cognate's multiple forms, possibly due to semantic shift or borrowing. This finding highlights that

accurately modeling the complexities of language can be even more crucial than simply using the most advanced phylogenetic techniques.

4.8.2 *Comparative mythology*

The study of myths is one field of application of phylogenetic analyses. In
a series of works, d'Huy applied phylogenetic approaches to a broad
range of problems, furnishing innovative contributions to a wide range of
questions using maximum parsimony, ML, and distance methods to compare results using different approaches. Let us mention his contributions
to the possible origin of some Greek myths (Polyphemus, Pygmalion) and
their relations to other traditions (d'Huy, 2013a, 2013b, 2023) or on the
distribution of matriarchy (d'Huy, 2023).

    The field of comparative mythology is developing steadily in many
directions: data collection and interpretation (Berezkin, 2015, 2017),
methodology, and data analytics (Thuillard *et al.*, 2018, 2021; Thuillard
and Le Quellec, 2017; Le Quellec and Sergent, 2017). Comparative folk-
loristics and mythology are two approaches that cannot be separated.
Similar mythic motifs may belong to the mythological corpus in some
traditions and be part of fairy tales in other traditions. Motifs related to the
Mesopotamian Etana myth about an eagle helping Etana to find the fertil-
ity plant in the sky are also recorded in many stories and fairy tales around
the Caucasus and the steppes (Thuillard, 2023a, 2003b). In recent years,
it has become increasingly clear that traditional phylogenetic trees, while
useful, cannot fully capture the complexities of cultural evolution. Myths,
folktales, and languages are often influenced by borrowing processes that
are not adequately represented by a simple branching tree structure.
Phylogenetic networks can better accommodate these complexities and
accurately describe how they evolve and spread across cultures. For
instance, Martini's (2020) analysis of the Cinderella story demonstrated
the power of networks to reveal the intricate patterns of borrowing and
hybridization that have shaped this tale across different cultures (see
Chapter 5 for a broader discussion).

4.8.3 *Archaeology*

In archaeology, phylogenetic approaches have a long history and are
reviewed in several articles or books (Rivero, 2016; Straffon, 2016;

Matzig *et al.*, 2023). It started with Petrie's (1917) work on seriation, which is, in modern terms, an approach corresponding to finding the order best fulfilling the consecutive-ones condition, a necessary step in constructing a phylogenetic network. Let us give a few examples.

In archaeology, much information can be gained from studying the chaîne opératoire, a French expression corresponding to a series of actions that transformed raw material into an artifact, for instance, a ceramic. Learning a potter's skill or making a carpet requires much investment and some willingness to transmit knowledge over generations. For example, the skills to make carpets are transmitted within a tribe among Iranian tribal populations and evolve mostly within the community (Tehrani and Collard, 2009). This process can be modeled as a phylogenetic tree. Phylogenetic trees correctly model the transmission of a chaîne opératoire in several examples from the European Middle Bronze Age (Manem, 2000) and the Chalcolithic of the Southern Levant (Roux, 2019). The phylogenic method explains the diversification of lithic projectile points in terms of design. The tree obtained by a maximum parsimony approach represents correlated changes in parameters (weight, shape, form) from one design to the others. The tree does not represent an evolution in time or space but the proximity of designs (O'Brien *et al.*, 2014; Cardillo and Alberti, 2023).

Phylogenetic trees can also describe complex occupation patterns. Prentiss *et al.* (2021) analyzed the long-term occupation and cultural evolution of house groups in an archaeological site in British Columbia over many generations about 1000 years ago. The phylogenetic analysis reveals patterns of transmission and innovation within the house pits. Despite the low number of characters, the data shows many clear bipartitions and a low deviation from a perfect phylogeny.

The latest two applications deal with morphometric data. Let us note that a phylogeny on morphometric data can be validated by verifying for each character how well the Kalmanson inequalities are fulfilled (see Section 4.2) and identifying each character's explanatory power (Thuillard and Fraix-Burnet, 2009).

4.8.4 *Cross-cultural studies*

Cross-cultural studies utilize phylogenetic methods to explore the relationships between cultural traits, languages, and genetics. A common approach consists of starting from a language or a genetic tree and probing whether some traits fit the tree. Greenhill *et al.* (2023) analyzed the Uto-Aztecan language family with a covarion model and a (relaxed-) clock to describe cognate evolution. They rooted the Uto-Aztecan tree around 4100 years ago near South California. The geographic origin is suggested based on the cultural practices, the language tree, and the likely diffusion of populations. The study further suggests, using root state reconstruction techniques, that gathering was the probable subsistence strategy. In another study, Watts *et al.* (2016) performed an ancestral state reconstruction on the variable "headhunting" among Austronesian traditions, suggesting that headhunting was practiced in proto-Austronesian culture. While these approaches offer valuable insights, reconstructing the past based on present-day data can be challenging. For example, the reconstruction of geographic origins relies on the assumption that language expansion is slow and gradual, which is not always true. Indeed, Neureiter *et al.* (2021) have shown with simulations that the reconstruction of the geographic origin is reliable if the language expansion is slow and gradual, which assumptions are generally difficult to verify. Therefore, the likelihood of the geographic origin is difficult to validate in the absence of indices and information permitting rooting or at least approximately positioning some nodes or edges. For that reason, studies are generally complemented with supplementary information from other fields.

The above approaches are ambitious, as they aim at reconstructing the past based mostly on recent information. It is exciting, but its potential and limits are still debated (Evans *et al.*, 2021). This type of study brings interesting suggestions that generally form a coherent narrative. Cultural studies bring a broad view of the past that can be further discussed with specialists in their field and confronted with other evidence. While there is some correlation between languages and genetic trees, there are also profound differences, the more in the past one goes. As an illustration, recent genetic studies on Bantu populations (Fortes-Lima *et al.*, 2024) have revealed complex migration routes, including multiple waves of migration and admixture with local populations. The authors caution against relying solely on modern language data to trace Bantu dispersion because of genetic admixture between linguistically distant populations (Koile *et al.*, 2022). The extent to which a trait may fit a genetic or linguistic tree depends on the cultural trait. In some cases, a trait might be unrelated: global musical diversity is largely independent of linguistic and genetic histories (Passmore *et al.*, 2024).

A second and different application consists of de-correlating data from a phylogenetic tree. In comparative studies, cultural traits are usually not independent. Closely related taxa on a phylogeny tend to share traits due to their common ancestry, and phylogenetic approaches permit removing that dependency (Freckleton *et al.* 2002). Studies deal with broad and diverse topics, such as the relationship between dwelling size and post-marital residence in agricultural societies (Hrnčíř *et al.* 2020) or polygyny. Currie *et al.* (2010) analyzed political complexity changes in Southeast Asia and the Pacific islands. The best-fitting model shows that political complexity rose and felt in small steps. Atkinson *et al.* (2016) analyzed deforestation, taking the phylogenetic tree for Austronesian languages to model the dependencies between regions. After eliminating the influence of phylogeny, the tendency of a man to have several wives is correlated, according to a study, to pathogen stress and assault frequency (Minocher *et al.*, 2019). Shcherbakova *et al.* (2024) show how different aspects of grammar can co-evolve over time. They find that the position of the verb in a sentence is a crucial factor in developing case systems and word order flexibility. These studies show the great diversity of addressed questions.

Several very performant and versatile software packages allow data processing. Let us mention PAUP (Swofford, 1993). for ML and parsimony, caper (Orme *et al.*, 2013), Phytools (Revell, 2012), BEAST (Suchard *et al.*, 2018), Mr Bayes (Ronquist and Huelsenbeck, 2003), and RevBayes (Höhna *et al.*, 2016). for Bayesian approaches. Mesquite is often used on continuous data (Maddison and Maddison, 2023; Lee *et al.*, 2006). Overall, the different packages offer a large collection of useful packages.

4.9 Outlook: Graph Neural Networks for Phylogenies

Graph neural networks can learn patterns and relationships in data structured on graphs. Graph neural networks have many applications (social network analysis, recommenders). They are also routinely used to build knowledge graphs that organize and retrieve information more easily. A graph neural network based on the GraphSage architecture was implemented in recommender applications like Uber Eat and Pinterest (Hamilton *et al.*, 2017; Rolim, 2022). The GraphSage architecture allows supervised and unsupervised learning. It implements a masking strategy similar to the one used for transformers in large language models during learning. The great strength of the network is that after learning, the model can classify new nodes based on the neighbor nodes without having to retrain the complete neural network. Attempts to use Graph Neural Networks or other neural network architectures for phylogenetic trees have led to some successes (Mo *et al.*, 2024), for instance, on molecular data alignments (Dotan *et al.*, 2023). It is too early to discuss the possible impact of neural networks on phylogenies in or outside of biology.

4.10 Complement: Continuous State Characters

The Wright–Fisher model describes genetic drift using a population of constant size N. In that model, each child has only one parent. At each generation, N samples are randomly drawn from an infinite population. Each parent transmits one of two possible character states (alleles) to children. The model describes the evolution of the expected frequency of each population defined by a character state. After a time dependent on the sample size, the whole population has a single character state corresponding to one of the two states. The model represents the evolution in a neutral framework without interaction between the different elements in the population and no evolutionary pressure.

The coalescent model is a mathematical model used to study the evolution of genetic diversity in a population. It describes how genetic lineages merge over time, eventually leading to a single common ancestor. The coalescent model is a continuous time approximation of the discrete Wright-Fisher model (Felsenstein, 1981). The model assumes constant population size. The expected time between coalescence events is inversely proportional to the size of the population. Smaller populations tend to coalesce more quickly. Cavalli-Sforza and Edwards (1967) showed that a Brownian motion model approximates the coalescence process well. The Brownian motion and the Gaussian distribution are at the base of the population evolution modeling.

Robert Brown was a botanist who lived during Darwin's time. His studies on the structure of the cell nucleus are quite well-known. Nowadays, he is mainly remembered for the Brownian motion. In an experiment, he observed the erratic movement of a pollen grain in water. In a famous paper, Einstein (1905) showed that assuming the movement is created by molecular movement, the displacement of the grain could be modeled using a Gaussian of mean zero. Most importantly, Einstein did relate the movement to the diffusion equation:

$$\partial u / \partial t = -D\left(\frac{\partial^2 u}{\partial x^2}\right). \tag{4.19}$$

The equation is central to the heat transfer and diffusion process. More important here is that it describes population dynamics. In a one-dimensional space, the solution of the diffusion equation is a Gaussian of the form:

$$u(x,t) = K \exp\left(-\frac{x^2}{4Dt}\right). \tag{4.20}$$

One observes that the variance increases linearly with time. As we will see below, this property permits the association of branch length with the variance.

After briefly explaining the rationality behind using Brownian motion to model continuous state characters, let us refine the model by integrating the possibility of branching. Figure 4.26 shows an example of a Brownian motion that models a continuous character, for instance, the allele frequency (an allele is one of several gene versions) in large populations. At the point marked by the arrow in Fig. 4.26, the population splits into two populations that independently evolve into populations B and C. During their common path, the two populations move away from their origin by some distance dA. After the paths separate, the two populations move away

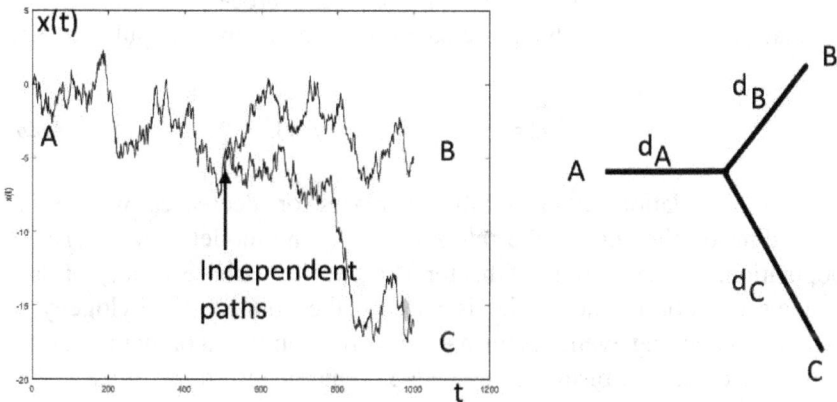

Fig. 4.26. Continuous variables (or characters) described by a coalescing Brownian process have a distance-based tree structure.

by *dB* and *dC*. Considering a very large population, one may define the distance *dA* using the covariance $Cov(X_B, X_C)$. The covariance of two independent processes equals zero. After some simple manipulations, one has

$$dA = Cov(X_B, X_C) = Var(X_A). \qquad (4.21)$$

In a tree, the allele distribution relates to the covariance matrix, which has a distance matrix structure. Mesquite is a much-used software for building trees from continuous characters (Maddison and Maddison, 2023).

4.11 Complement: The Birth and Death Model of Evolution

The Birth and Death model is another important model of evolution. The model relates the growth of a population to two parameters describing the probability of birth μ and death λ. Birth events correspond to the appearance of new species. In the model, new species are added at a rate μ and disappear at a rate l. In its simplest form, a birth-death model is of the form:

$$P_{i,i+1}(\Delta t) \cong \lambda \Delta t,$$
$$P_{i,i-1}(\Delta t) \cong \mu \, \Delta t, \qquad (4.22)$$
$$P_{i,i}(\Delta t) \cong 1 - (\lambda + \mu)\Delta t,$$

$P_{i,i+1}, P_{i,i-1}$, are the probability of a birth or a death process within an infinitesimal time interval. Solving the equation, one obtains a population size $N(t)$:

$$N(t) = N(0) \, \exp((\mu - \mu)t). \qquad (4.23)$$

The population exponentially increases or decreases with time, depending on the sign of the rate $r = \lambda - \mu$. The model may describe a population, the evolution of bacteria, a species' arborescence, or language's evolution. The model is used in the context of phylogeny to describe branching events. A birth event corresponds to a branching event. A node is created at birth, and two edges "emerge" from the process. The disappearance of a branch with time corresponds to a death event. In this model, the probability of branching increases exponentially with the number of branches. When applied to evolution, the model's main challenge is quantifying the birth and death rates from partial observations. For instance, fossil records furnish an incomplete picture of evolution, with many marine species over-represented due to their larger fossilization probability than land animals (Stadler, 2009).

References

Atkinson, Q. D., Coomber, T., Passmore, S., Greenhill, S. J., & Kushnick, G. (2016). Cultural and environmental predictors of pre-European deforestation on Pacific Islands, *PLoS One*, 11(5), p. e0156340.

Barbançon, F., Evans, S. N., Nakhleh, L., Ringe, D., & Warnow, T. (2013). An experimental study comparing linguistic phylogenetic reconstruction methods, *Diachronica, International Journal for Historical Linguistics*, 30(2), pp. 143–170.

Barbrook, A. C., Howe, C. J., Blake, N., & Robinson, P. (1998). The phylogeny of the Canterbury Tales, *Nature*, 394(6696), p. 839.

Barido-Sottani, J., Pohle, A., De Baets, K., Murdock, D., & Warnock, R. C. (2023). Putting the F into FBD analysis: Tree constraints or morphological data? *Paleontology*, 66(6), p. e12679.

Berezkin, Y. (2015). Folklore and mythology catalogue: Its lay-out and potential for research. Frog & Lukin, K. (eds.), *Between Text and Practice: Mythology, Religion and Research. A special issue of RMN Newsletter*. Helsinki: University of Helsinki, pp. 58–70. Available at https://www.academia.edu/14481230/ last accessed on 20 October 2024.

Berezkin, Y. (2017). Maths meets myths: Quantitative approaches to ancient narratives. Understanding complex systems, Kenna R., & M. MacCarron M., MacCarron, P. (eds.), *Peopling of the New World from Data on Distributions of Folklore Motifs*, (Springer, Cham.), pp. 71–89.

Borges, R., Machado, J. P., Gomes, C., Rocha, A. P., & Antunes, A. (2019). Measuring phylogenetic signal between categorical traits and phylogenies, *Bioinformatics*, 35(11), pp. 1862–1869.

Bouckaert, R. R., & Robbeets, M. (2017). Pseudo Dollo models for the evolution of binary characters along a tree, BioRxiv, 207571.

Bryant, D., & Moulton, V. (2004). Neighbor-net: An agglomerative method for constructing phylogenetic networks, *Molecular Biology and Evolution*, 21(2), pp. 255–265.

Canby, M. E., Evans, S. N., Ringe, D., & Warnow, T. (2024). Addressing polymorphism in linguistic phylogenetics, *Transactions of the Philological Society*, 122, pp. 191–222.

Cardillo, M., & Alberti, J. (2023). Explaining the diversification of lithic projectile points from the northern Patagonian coast (Argentina) during the

Holocene using phylogenetic and comparative methods, *Journal of Archaeological Science: Reports*, 51, p. 104144.

Cavalli-Sforza, L. L., & Edwards, A. W. (1967). Phylogenetic analysis. Models and estimation procedures, *American Journal of Human Genetics*, 19(3 Pt 1), pp. 233–257.

Currie, T. E., Greenhill, S. J., Gray, R. D., Hasegawa, T., & Mace, R. (2010). Rise and fall of political complexity in island Southeast Asia and the Pacific, *Nature*, 467(7317), pp. 801–804.

d'Huy, J., Le Quellec, J. L., Thuillard, M., Berezkin, Y. E., Lajoye, P., & Oda, J. I. (2023). Little statisticians in the forest of tales: Towards a new comparative mythology, *Fabula*, 64(1–2), pp. 44–63.

d'Huy, J. (2013a). Polyphemus (Aa. Th. 1137): A phylogenetic reconstruction of a prehistoric tale. *Nouvelle Mythologie Comparée/New Comparative Mythology*, 1(1), pp. 3–18.

d'Huy, J. (2013b). Il y a plus de 2000 ans, le mythe de Pygmalion existait en Afrique du nord. *Préhistoires méditerranéennes*, 4, (in French).

d'Huy, J. (2023). *Cosmogonies: la Préhistoire des Mythes*, (La Découverte) (in French).

Dotan, E., Belinkov, Y., Avram, O., Wygoda, E., Ecker, N., Alburquerque, M., ... & Pupko, T. (2023). Multiple sequence alignment as a sequence-to-sequence learning problem, *Eleventh International Conference on Learning Representations (ICLR 2023)*.

Einstein, A. (1905). On the motion of small particles suspended in liquids at rest required by the molecular-kinetic theory of heat, *Annalen der Physik*, 17(208), pp. 549–560.

Evans, C. L., Greenhill, S. J., Watts, J., List, J. M., Botero, C. A., Gray, R. D., & Kirby, K. R. (2021). The uses and abuses of tree thinking in cultural evolution, *Philosophical Transactions of the Royal Society B*, 376(1828), https://doi.org/10.1098/rstb.2020.0056.

Farris, J. S. (1989). The retention index and the rescaled consistency index, *Cladistics*, 5(4), pp. 417–419.

Felsenstein, J. (1981). Evolutionary trees from gene frequencies and quantitative characters: finding maximum likelihood estimates, *Evolution*, pp. 1229–1242.

Fischer, M. (2023). Defining binary phylogenetic trees using parsimony, *Annals Combinatorics*, 27(3), pp. 457–467.

Fitch, W. M. (1970). Distinguishing homologous from analogous proteins, *Systematic Zoology*, 19(2), pp. 99–113.

Fortes-Lima, C. A., Burgarella, C., Hammarén, R., Eriksson, A., Vicente, M., Jolly, C., ... & Schlebusch, C. M. (2024). The genetic legacy of the expansion of Bantu-speaking peoples in Africa, *Nature*, 625(7995), pp. 540–547.

Freckleton, R. P., Harvey, P. H., & Pagel, M. (2002). Phylogenetic analysis and comparative data: A test and review of evidence, *The American Naturalist*, 160(6), pp. 712–726.

Gascuel, O., & Steel, M. (2006). Neighbor-joining revealed, *Molecular Biology and Evolution*, 23(11), pp. 1997–2000.

Gray, R. D., & Atkinson, Q. D. (2003). Language-tree divergence times support the Anatolian theory of Indo-European origin, *Nature*, 426(6965), pp. 435–439.

Greenhill, S. J., Currie, T. E., & Gray, R. D. (2009). Does horizontal transmission invalidate cultural phylogenies? *Proceedings of the Royal Society B: Biological Sciences*, 276(1665), pp. 2299–2306.

Greenhill, S. J., Haynie, H. J., Ross, R. M., Chira, A. M., List, J. M., Campbell, L., ... & Gray, R. D. (2023). A recent northern origin for the Uto-Aztecan family, *Language*, 99(1), pp. 81–107.

González-José, R., Escapa, I., Neves, W. A., Cúneo, R., & Pucciarelli, H. M. (2008). Cladistic analysis of continuous modularized traits provides phylogenetic signals in Homo evolution, *Nature*, 453(7196), pp. 775–778.

Grimm, J., & Grimm, W. (1884). *Grimm's Household Tales: With the Author's Notes*, (Vol. 2) (G. Bell).

Harmon, L. (2019). *Phylogenetic Comparative Methods*. Self-edition online: CreateSpace Independent Publishing Platform, Scotts Valley.

Heath, T. A., Huelsenbeck, J. P., & Stadler, T. (2014). The fossilized birth-death process for coherent calibration of divergence-time estimates, *Proceedings of the National Academy of Sciences*, 111(29), pp. E2957–E2966.

Heggarty, P., Anderson, C., Scarborough, M., King, B., Bouckaert, R., Jocz, L., ... & Gray, R. D. (2023). Language trees with sampled ancestors support a hybrid model for the origin of Indo-European languages, *Science*, 381(6656), p. eabg0818.

Höhna, S., Landis, M. J., Heath, T. A., Boussau, B., Lartillot, N., Moore, B. R., ... & Ronquist, F. (2016). RevBayes: Bayesian phylogenetic inference using graphical models and an interactive model-specification language, *Systematic Biology*, 65(4), pp. 726–736.

Howe, C. J., & Windram, H. F. (2011). Phylomemetics — Evolutionary analysis beyond the gene, *PLoS Biology*, 9(5), p. e1001069.

Howe, C. J., & Windram, H. F. (2023), Manuscript traditions. J. Tehrani *et al.* (eds.), *The Oxford Handbook of Cultural Evolution*.

Hrnčíř, V., Duda, P., Šaffa, G., Květina, P., & Zrzavý, J. (2020). Identifying post-marital residence patterns in prehistory: A phylogenetic comparative analysis of dwelling size, *PLoS One*, 15(2), p. e0229363.

Hyytiäinen, P. (2022). The changing text of acts: A phylogenetic approach, *TC: A Journal of Biblical Textual Criticism*, 26, pp. 1–28.

Jukes, T. H., & Cantor, C. R. (1969). Evolution of protein molecules, *Mammalian Protein Metabolism*, 3(24), pp. 21–132.

Kalmanson, K. (1975). Edgeconvex circuits and the traveling salesman problem, *Canadian Journal Mathematics*, 27(5), pp. 1000–1010.

Kanojia, D., Dubey, A., Kulkarni, M., Bhattacharyya, P., & Haffari, G. (2019). Utilizing word embeddings based features for phylogenetic tree generation of Sanskrit texts, *Proceedings of the 6th International Sanskrit Computational Linguistics Symposium*, pp. 152–165.

Koile, E., Greenhill, S. J., Blasi, D. E., Bouckaert, R., & Gray, R. D. (2022). Phylogeographic analysis of the Bantu language expansion supports a rainforest route, *Proceedings of the National Academy of Sciences*, 119(32), p. e2112853119.

Kozlov, A. M., Darriba, D., Flouri, T., Morel, B., & Stamatakis, A. (2019). RAxML-NG: A fast, scalable and user-friendly tool for maximum likelihood phylogenetic inference, *Bioinformatics*, 35(21), pp. 4453–4455.

Le Quellec, J. L., & Sergent, B. (2017) *Dictionnaire Critique de Mythologie* (CNRS éditions).

Lee, C., Blay, S., Mooers, A. Ø., Singh, A., & Oakley, T. H. (2006). CoMET: A Mesquite package for comparing models of continuous character evolution on phylogenies, *Evolutionary Bioinformatics*, https://doi.org/10.1177/117693430600200022.

Levy, D., & Pachter, L. (2011). The neighbor-net algorithm, *Advances in Applied Mathematics*, 47(2), pp. 240–258.

Lewis, P. O. (2001). A likelihood approach to estimating phylogeny from discrete morphological character data, *Systematic Biology*, 50(6), pp. 913–925.

Maddison, W. P., & Maddison, D. R. (2023). Mesquite: A modular system for evolutionary analysis. Version 3.81 http://www.mesquiteproject.org.

Manem, S. (2020). Modeling the evolution of ceramic traditions through a phylogenetic analysis of the Chaînes Opératoires: The European Bronze Age as a case study, *Journal of Archaeological Method and Theory*, 27(4), pp. 992–1039.

Marmerola, G. D., Oikawa, M. A., Dias, Z., Goldenstein, S., & Rocha, A. (2016). On the reconstruction of text phylogeny trees: Evaluation and analysis of textual relationships, *PLoS One*, 11(12), p. e0167822.

Martini, G. (2020). *Cinderella: An Evolutionary Approach to the Study of Folktales* (Doctoral dissertation, Durham University).

Matzig, D. N., Schmid, C., & Riede, F. (2023). Mapping the field of cultural evolutionary theory and methods in archaeology using bibliometric methods, *Humanities and Social Sciences Communications*, 10(1), pp. 1–17.

McCollum, J., & Turnbull, R. (2024). Using Bayesian phylogenetics to infer manuscript transmission history, *Digital Scholarship in the Humanities*, 39(1), pp. 258–279.

Minocher, R., Duda, P., & Jaeggi, A. V. (2019). Explaining marriage patterns in a globally representative sample through socio-ecology and population history: A Bayesian phylogenetic analysis using a new supertree, *Evolution and Human Behavior*, 40(2), pp. 176–187.

Mo, Y. K., Hahn, M. W., & Smith, M. L. (2024). Applications of machine learning in phylogenetics, *Molecular Phylogenetics and Evolution*, 196, p. 108066.

Nakhleh, L., Ringe, D. A., & Warnow, T. (2005). Perfect phylogenetic networks: A new methodology for reconstructing the evolutionary history of natural languages, *Language*, 81(2), pp. 382–420.

Neureiter, N., Ranacher, P., van Gijn, R., Bickel, B., & Weibel, R. (2021). Can Bayesian phylogeography reconstruct migrations and expansions in linguistic evolution? *Royal Society Open Science*, 8(1), p. 201079.

Ng, S. K., Krishnan, T., & McLachlan, G. J. (2012). The EM algorithm. *Handbook of Computational Statistics: Concepts and Methods*, pp. 139–172.

O'Brien, M. J., Boulanger, M. T., Buchanan, B., Collard, M., Lyman, R. L., & Darwent, J. (2014). Innovation and cultural transmission in the American Paleolithic: Phylogenetic analysis of eastern Paleoindian projectile-point classes, *Journal of Anthropological Archaeology*, 34, pp. 100–119.

Orme, D., Freckleton, R., Thomas, G., Petzoldt, T., Fritz, S., Isaac, N., & Pearse, W. (2013). The Caper package: Comparative analysis of phylogenetics and evolution in R. *Methods in Ecology and Evolution*, 3, pp. 145–151.

Oteo-Garcia, G., Oteo, J. A. (2021). A geometrical framework for f-statistics. *Bulletin of Mathematical Biology*, 83(2), p. 14.

Passmore, S., Wood, A. L., Barbieri, C., Shilton, D., Daikoku, H., Atkinson, Q. D., & Savage, P. E. (2024). Global musical diversity is largely independent of linguistic and genetic histories, *Nature Communications*, 15(1), p. 3964.

Patterson, N., Moorjani, P., Luo, Y., Mallick, S., Rohland, N., Zhan, Y., ... & Reich, D. (2012). Ancient admixture in human history, *Genetics*, 192(3), pp. 1065–1093.

Pellard, T., Ryder, R., & Jacques, G. (2024). *The Family Tree model*. The Wiley Blackwell Companion to Diachronic Linguistics.

Petrie, W. F. (1917). *Prehistoric Egypt* (Vol. 9) (Oxbow Books).

Posth, C., Yu, H., Ghalichi, A., Rougier, H., Crevecoeur, I., Huang, Y., ... & Krause, J. (2023). Palaeogenomics of upper paleolithic to neolithic European hunter-gatherers, *Nature*, 615(7950), pp. 117–126.

Prentiss, A. M., Walsh, M. J., Foor, T. A., Hampton, A., & Ryan, E. (2020). Evolutionary household archaeology: Inter-generational cultural transmission at housepit 54, Bridge River site, *British Columbia. Journal of Archaeological Science*, 124, p. 105260.

Revell, L. J. (2012). Phytools: An R package for phylogenetic comparative biology (and other things), *Methods in Ecology and Evolution*, (2), pp. 217–223.

Ribeiro, D., Borges, R., Rocha, A. P., & Antunes, A. (2023). Testing phylogenetic signal with categorical traits and tree uncertainty, *Bioinformatics*, 39(7), p. btad433.

Ringe, D., Warnow, T., & Taylor, A. (2002). Indo-European and computational cladistics, *Transactions of the Philological Society*, 100(1), pp. 59–129.

Rivero, D. G. (2016). Darwinian archaeology and cultural phylogenetics, *Cultural Phylogenetics: Concepts and Applications in Archaeology*, pp. 43–72.

Rolim, L. L. (2022). *Embedding of Bipartite Graphs via Neural Networks with Application to User-Item Recommendations* (Doctoral dissertation, Universidade Federal do Rio de Janeiro).

Ronquist, F., & J. P. Huelsenbeck. (2003). MRBAYES 3: Bayesian phylogenetic inference under mixed models, *Bioinformatics*, 19, pp. 1572–1574.

Ronquist, F., Klopfstein, S., Vilhelmsen, L., Schulmeister, S., Murray, D. L., & Rasnitsyn, A. P. (2012). A total-evidence approach to dating with fossils, applied to the early radiation of the Hymenoptera, *Systematic Biology*, 61(6), pp. 973–999.

Roux, V. (2019). The Ghassulian ceramic tradition: A single chaîne opératoire prevalent throughout the Southern Levant, *Journal of Eastern Mediterranean Archaeology & Heritage Studies*, 7(1), pp. 23–43.

Saitou, N., & Nei, M. (1987). The neighbor-joining method: A new method for reconstructing phylogenetic trees, *Molecular Biology and Evolution*, 4(4), pp. 406–425.

Sankoff, D. (1985). Simultaneous solution of the RNA folding, alignment and protosequence problems, *SIAM Journal on Applied Mathematics*, 45(5), pp. 810–825.

Shapiro, B., Rambaut, A., & Drummond, A. J. (2006). Choosing appropriate substitution models for the phylogenetic analysis of protein-coding sequences, *Molecular Biology and Evolution*, 23(1), pp. 7–9.

Semple, C., & Steel, M. (2002). Tree reconstruction from multistate characters, *Advances in Applied Mathematics*, 28(2), pp. 169–184.

Semple, C., & Steel, M. (2003). *Phylogenetics*, (Vol. 24) (Oxford University Press).

Shcherbakova, O., Blasi, D. E., Gast, V., Skirgård, H., Gray, R. D., & Greenhill, S. J. (2024). The evolutionary dynamics of how languages signal who does what to whom, *Scientific Reports*, 14(1), p. 7259.

Stadler, T. (2009). On incomplete sampling under birth-death models and connections to the sampling-based coalescent, *Journal of Theoretical Biology*, 261(1), pp. 58–66.

Stevens, K., & Gusfield, D. (2010). Reducing multistate to binary perfect phylogeny with applications to missing, removable, inserted, and deleted data, *Algorithms in Bioinformatics: 10th International Workshop, WABI 2010*, Liverpool, UK, September 6–8, 2010, (Springer, Berlin Heidelberg), pp. 274–287.

Straffon, L. M. (ed.). (2016). *Cultural Phylogenetics: Concepts and Applications in Archaeology*, (Vol. 4) (Springer).

Suchard, M. A., Lemey, P., Baele, G., Ayres, D. L., Drummond, A. J., & Rambaut, A. (2018). Bayesian phylogenetic and phylodynamic data integration using BEAST 1.10, *Virus Evolution*, 4, p. vey016.

Swofford, D. L. (1993). PAUP, phylogenetic analysis using parsimony. Version 3.1. Computer program distributed by the Illinois Natural History Survey.

Tehrani, J. J. (2013). The phylogeny of Little Red Riding Hood, *PLoS One*, 8(11), p. e78871.

Tehrani, J., & Collard, M. (2002). Investigating cultural evolution through biological phylogenetic analyses of Turkmen textiles, *Journal of Anthropological Archaeology*, 21(4), pp. 443–463.

Thuillard, M. (2007). Minimizing contradictions on circular order of phylogenic trees, *Evolutionary Bioinformatics*, 3, pp. 267–277.

Thuillard, M. (2008). Minimum contradiction matrices in whole genome phylogenies, *Evolutionary Bioinformatics*, 4, EBO-S909, pp. 237–247.

Thuillard, M. (2009). Evolutionary biology: Concept, modeling, and application. Pontarotti, P. (ed.), *Why Phylogenetic Trees are Often Quite Robust Against Lateral Transfers?* (Springer), pp. 269–283.

Thuillard, M. (2021). Analysis of the worldwide distribution of the 'Man or Animal in the Moon' motifs, *Folklore: Electronic Journal of Folklore*, 84, pp. 127–144.

Thuillard, M. (2023a). *Wavelet in Soft Computing*, 2nd edn. (World Scientific).

Thuillard, M. (2023b). *Le Gai Sçavoir: Mélanges en Hommage à Jean-Loïc Le Quellec*. d'Huy, J., Lajoye, P., & Duquesnoy, F., Le Dessous des Cartes (eds.): Mythologie Comparée (Archeone) (in French).

Thuillard, M., & Fraix-Burnet, D. (2009). Phylogenetic applications of the minimum contradiction approach on continuous characters, *Evolutionary Bioinformatics*, 5, EBO-S2505, pp. 33–46.

Thuillard, M., & Fraix-Burnet, D. (2015). Phylogenetic trees and networks reduce to phylogenies on binary states: Does it furnish an explanation to the robustness of phylogenetic trees against lateral transfers? *Evolutionary Bioinformatics*, 11, EBO-S28158, pp. 213–221.

Thuillard, M., & Le Quellec, J. L. (2017). A phylogenetic interpretation of the canonical formula of myths by Lévi-Strauss, *Cultural Anthropology and Ethnosemiotics*, 3(2), pp. 1–12.

Thuillard, M., Le Quellec, J.-L., d'Huy, J., & Berezkin, Y. (2018). A large-scale study of world myths, *Trames: Journal of the Humanities and Social Sciences*, 22(4), pp. 407–424.

Watts, J., Sheehan, O., Greenhill, S. J., Gomes-Ng, S., Atkinson, Q. D., Bulbulia, J., & Gray, R. D. (2015). Pulotu: Database of Austronesian supernatural beliefs and practices, *PLoS One*, 10(9), p. e0136783.

Windram, H. F., Charlston, T., Tomita, Y., & Howe, C. J. (2022). A phylogenetic analysis of two preludes from JS Bach's Well-Tempered Clavier II, *Early Music*, 50(3), pp. 373–393.

Zhang, C., Stadler, T., Klopfstein, S., Heath, T. A., & Ronquist, F. (2016). Total-evidence dating under the fossilized birth-death process, *Systematic Biology*, 65(2), pp. 228–249.

Chapter 5

Applications of Phylogenetic Networks

5.1 Introduction

In Chapter 4, we introduced phylogenetic trees and laid the basis for phylogenetic networks and their validation in the context of binary-state characters. In this chapter, we want to complete the presentation and discuss some opportunities and issues with phylogenetic networks. A phylogenetic tree is a too-crude representation of the relationships between taxa (i.e., end nodes) in many applications with horizontal transfers. Lateral (or horizontal) transfer describes the transmission of characters between taxa without a direct vertical inheritance from parents to their offspring. Deviations from a tree topology are often represented by a line or an arrow (in rooted trees) relating two nodes. This representation preserves roughly a tree structure, and the lines or arrows indicate regions deviating from a tree. It is the most intuitive form of reticular structures. Numerous methods exist for inferring phylogenetic networks using maximum parsimony, maximum likelihood, or even Bayesian approaches. Many software packages for rooted (Dendroscope; Huson and Scornavacca, 2012) or unrooted phylogenetic networks are available (Huson *et al.*, 2010; Boc *et al.*, 2012; Gusfield, 2014) with applications mostly in biology (Huson and Bryant, 2006).

Phylogenetic trees are widely used outside of biology to describe the evolution of some cultural traits. While most researchers agree on the importance of networks in furnishing a more realistic picture of evolution, their application is still limited. One of the main reasons is that the description of data with a phylogenetic network requires prohibitive computing power in many practical applications. The exception is NeighborNet (Bryant and Moulton, 2004), a distance-based algorithm that requires low computing power even for over 2000 taxa. Identifiability is another issue with phylogenetic networks. Non-identifiability in a model refers to the situation where it is impossible to uniquely determine the true values of the model's parameters. As complexity is added to a model to describe lateral transfers or hybridization events, the model becomes more versatile and capable of fitting a wider range of data patterns. The increased versatility can lead to identifiability issues, with very different parameter combinations producing comparable results. Some mild conditions on the data are sometimes sufficient to prevent the problem (Warnow *et al.*, 2024). As a rule of thumb, identifiability becomes an issue if many parameters must be added to a tree model to describe the data or if the model diverges much from a phylogenetic tree or network. In other words, phylogenetic networks are great tools when the deviations to a tree topology are not too large.

5.2　Applications of Phylogenetic Networks

Phylogenetic networks have applications in biology, cultural studies, and language evolution. One of the first applications of phylogenetic networks (Legendre and Makarenkov, 2002) was fitting the topology of a reticular structure on the Canadian Rivers map, which describes the dispersal of freshwater fish after the last glaciation.

Some of the most interesting studies combine synthetically different analytical methods with specialist knowledge. For example, Zhao *et al.* (2023) analyzed the spatiotemporal dynamics of bread wheat emergence and dispersal by examining the different wheat species. The bread wheat originated from the southwest coast of the Caspian Sea, spread across Eurasia, and reached Europe, South Asia, and East Asia ~7,000 to ~5,000 BP. Several subspecies originated through hybridization between expanding bread wheat and locally pre-existing wheat. Interspecific hybridization is common during the range expansion of species. PhyloNet (Wen *et al.*, 2018) inferred the reticulate phylogenetic networks describing the evolution of these subspecies based on orthologous genes. Combining the geographic information on locally pre-existing wheat and phylogenetic information on the different varieties with an Estimating Effective Migration Surfaces (EEMS) approach led to the identification of three dispersal routes connecting Central and East Asia. EMMS visualizes non-homogeneous gene flows on a geographic map (Petkova *et al.*, 2016), which is possible if the rate at which genetic similarity decays with distance is spatially heterogeneous (see also Section 5.4.3). This study of wheat dispersion efficiently combines different techniques using some redundancy. In the absence of redundancy, combining several methods requires that each one is reliable, and one must be aware that the errors of each method cumulate.

Outside of biology, phylogenetic networks using maximum parsimony, maximum likelihood, or a Bayesian approach have been implemented in the analysis of language evolution (Nakhleh *et al.*, 2005; Boc *et al.*, 2010; List *et al.*, 2014; Neureiter *et al.*, 2024). One is in the somewhat paradoxical situation that linguists criticize phylogenetic trees, as trees neglect important evolution mechanisms such as horizontal transfers, and phylogenetic networks are not well received because they are too complex. The situation is slowly changing with new studies and tools. Chinese dialects have a complex evolution, and the extra effort for developing phylogenetic networks is completely justified. List *et al.* (2014) showed that a tree model cannot readily explain most characters. More generally, one observes the emergence of specialized tools. Neureiter *et al.* (2024) developed an open phylogenetic network tool. In a case study, the authors identified known contact events in the history of Indo-European and loanwords.

5.3 Outer Planar Networks and Their Interpretations

A distance matrix fulfilling the Kalmanson inequalities fits exactly a perfect phylogenetic network. More precisely, it does fit an outer planar network. An outer planar network has no edge crossing, and all its end nodes lie on the outer face of the network. Figure 5.1 shows an example of an outer planar network.

As discussed in Chapter 4, a phylogenetic network perfectly represents binary character states fulfilling the consecutive-ones condition. There are two main interpretations of a phylogenetic network. A phylogenetic network can be interpreted as a natural extension of phylogenetic trees in DNA studies. A network is a tree modified by horizontal transfer. Such an interpretation is the one for DNA studies. The approach can lead to new classifications. In paleobotany, an analysis supported by

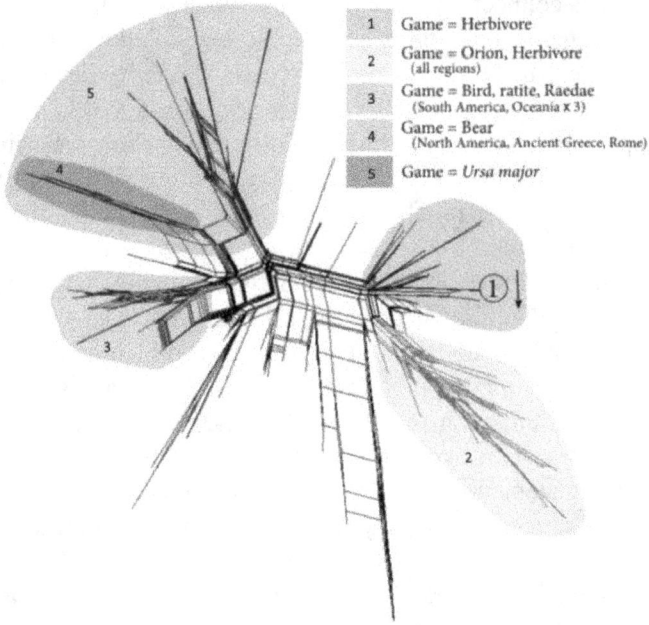

Fig. 5.1.   Example of an outer planar network. The network represents data characterizing the different versions of a myth family described as the cosmic hunt, in which catasterized humans and animals follow each other in the sky. Thuillard *et al.* (2018b).

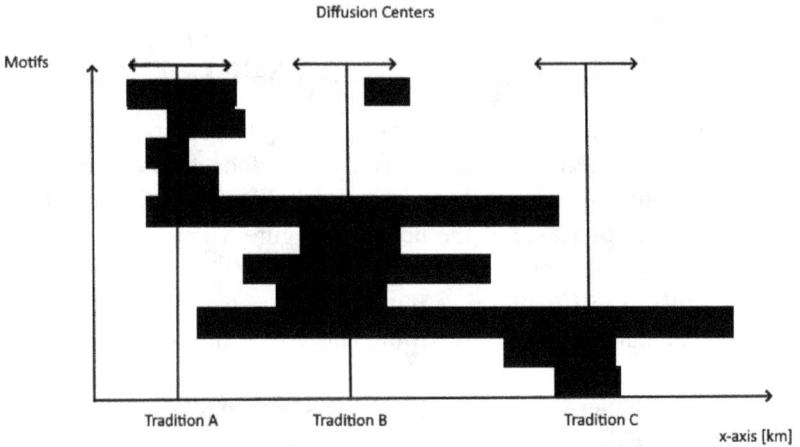

Fig. 5.2.   An outer planar network perfectly represents characters fulfilling the circular consecutive-ones conditions. In cultural studies, phylogenetic networks allow multiple centers of origin for the different characters. Applications of outer planar networks.

NeigborNet resulted in the classification revision of the fossil Osmundaceae (Bomfleur *et al.*, 2017; Coiro, 2024). A completely different interpretation of the same phylogenetic network in non-genetic applications is that it represents the diffusion of characters having multiple geographic origins (Fig. 5.2). If one considers the propagation of cultural elements, both interpretations are fundamentally different. Consider the evolution of folktales. Different stories appear in different cultures, and each may propagate to its neighbors. The network might reveal that a particular motif (e.g., a trickster story) arose in one region and then spread through cultural contact. Deviations to a phylogenetic network may indicate that a story appeared independently in several regions.

5.3.1 *Admixture trees for ancient DNA: F-statistics*

Ancient DNA analyses on whole genomes have deeply changed our comprehension of the past by tracing the migration patterns of different groups and showing the extent of admixture events in the genetic content of human populations. Applications of phylogenetic networks in describing ancient DNA are often based on summary statistics describing allele frequencies. The resulting rooted graphs have a general form familiar to users of rooted phylogenetic trees and, therefore, are quite intuitive to interpret. Analyzing the frequency of different single-nucleotide polymorphisms (SNPs) at a set of loci is one widespread method of studying admixtures. SNPs are variations on a single nucleotide associated with a position called locus on the gene. In whole genome analysis, data are generally described by an admixture graph, a tree augmented by admixture events. Finding the best admixture scenarios is computer-intensive and typically limited to a few taxa. The two main approaches are qpGraph (Patterson *et al.*, 2012) and Treemix (Pickrell and Pritchard, 2012). Both methods use allele frequency data as input. They differ in the detailed statistics but are not fundamentally different in their approach. The qpGraph approach starts from a small tree and adds taxa (or populations) one at a time, adding admixture events if required. The identification of admixture is supported by an F-statistics analysis (Lipson *et al.*, 2020; Patterson *et al.*, 2020). The F_2-statistics is computed with the following expression (as a convention, instances of F-statistics are represented by a small f).

$$f_2(i,j) = \sum_{s=1}^{S} (p_{i,s} - p_{j,s})^2, \tag{5.1}$$

where S the number of loci and $p_{i,s} = (q_1, q_2, \ldots, q_n)$, the probabilities of the n alleles for the ith taxon. An allele is a different version of the same gene, and a locus refers to a specific position on a chromosome where a particular gene or genetic marker is located.

The f_2 value relates to the difference between the two taxas. The f_2 value between two taxa is zero if, on each locus, the alleles of the two taxa have the same distribution. The f_3 value captures the relationship between two taxa with a third taxon, called the reference taxon.

The equation for the F_3-statistics:

$$f_3(i;j,k) = \sum_{s=1}^{S} (p_{i,s} - p_{j,s})(p_{i,s} - p_{k,s}) \tag{5.2}$$

From Eqs. (5.1) and (5.2), one verifies a useful relation between the f_3 and the f_2 statistics:

$$f_3(i;j,k) = \frac{1}{2}(f_2(i,j) + f_2(i,k) - f_2(j,k)). \tag{5.3}$$

Interestingly, the f_3-statistics is similar to Eq. (4.3) in a distance-based phylogenetic tree. Both equations are essentially measuring shared evolutionary history. In a distance-based tree, $Y_{i,j}^n$ corresponds to the shortest distance from a reference taxon n to the common path relating two taxa (i,j). The f_3 statistic measures the shared genetic drift of the two taxa from a third reference (The structure is identical to $Y_{j,k}^i$, but we did not change the indices to respect the conventions).

From the discussion in Section 3.5.2, the Kalmanson condition becomes $f_3(i;j,k) \geq f_3(i;j,l)$ with $i \leq j \leq k \leq l$ being successive in a circular order of the taxa on a tree or a network.

For distance-based phylogenetic trees, the circular order of the taxa is quite robust against lateral transfer (Thuillard, 2009). Lateral transfers between adjacent taxa preserve the Kalmanson inequalities (Thuillard and Fraix-Burnet, 2015). It is also the case for admixture graphs, provided admixtures are between adjacent taxa in a circular order (see Complement Section 5.4). So, if admixtures are between adjacent taxa, the f_3 matrix can be ordered to fulfill the Kalmanson inequalities with values never increasing away from the diagonal, provided the data fit a phylogenetic tree or network. In other words, the non-fulfillment of the inequality $f_3(i; j, k) \geq f_3(i; j, l)$ with $i \leq j \leq k \leq l$ being successive indicates a deviation from a phylogenetic network structure. (Let us note that a negative f_3 is associated with a deviation from a tree topology (Patterson *et al.*, 2012), while the inequality on f_3 is directly related to the f_4 statistics and deviations to a phylogenetic network structure.)

Let us look at one ancient DNA study using the f_3 statistic. Posth *et al.* (2023) investigated the genetic relationships between ancient hunter-gatherer groups from Portugal to Russia using genome-wide data from individuals associated with different archaeological cultures. They aimed to understand population structure and admixture patterns between the Upper Paleolithic and the Neolithic periods. The authors furnish the data on the f_3 statistics in the supplement to their articles. We have ordered the data using the minimum contradiction method in Chapter 3 and represented the ordered matrix in a heatmap. Figure 5.3 shows a subsection of the complete heatmap. A color scale codes the f_3 values. Large f_3 values are yellow, while very low values are coded blue.

The main result in Posth *et al.* (2023) can be read in the heatmap. The archaeologically determined Gravettian cultures divide genetically into two geographically distinct groups: Fournol (France, Germany, and Belgium) and Vestonice (Eastern regions, Italy). Further, one observes that the Iberian hunter-gatherer cluster and, to a lesser extent, the

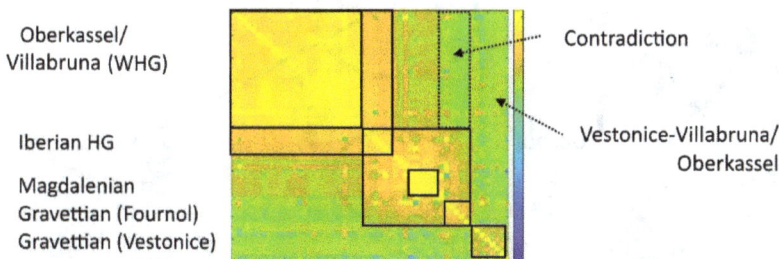

Fig. 5.3. Ordered f_3-matrix. The matrix shows a broader data view, focusing on the Gravettian and Magdalenian cultures. The f_3-values of the Iberian hunter-gatherers connect the Magdalenian and the Oberkassel/Villabruna cluster. One observes a contradiction between the Gravettian Vestonice, the Fournol, and the Vestonice cluster. The higher contribution of the Vestonice cluster than the Fournol cluster to the Oberkassel/Villanova cluster may be interpreted as being slightly closer to the Vestonice than the Fournol DNA.

Magdalenian cluster do not fit a tree. A deviation from the Kalmanson inequalities is observed with the Oberkassel/Villabruna and the Gravettian clusters, hinting at the Epigravettian Villabruna cluster being closer to Gravettian Vestonice than to the Fournol cluster. The ordered matrix does not fit well with a phylogenetic tree topology but does fit well with a phylogenetic network except for that one cluster.

The analysis of the f_3 matrix in the form of a heatmap is helpful during preprocessing to choose which taxa should be included in the admixture analysis. The level of contradiction in the order is also a good indication of the complexity of the admixture tree. As an exploratory tool, the ordering approach can guide the selection of the taxa for more advanced studies. In the absence of contradiction, one expects the order of the taxa on the admixture tree to be compatible with the ordered f_3-matrix and the admixture mostly between adjacent taxa or clusters. Analyzing published admixture trees (Maier *et al.*, 2023) has shown that some trees are very sensitive to the taxa choice. The comparison of the taxa order on the

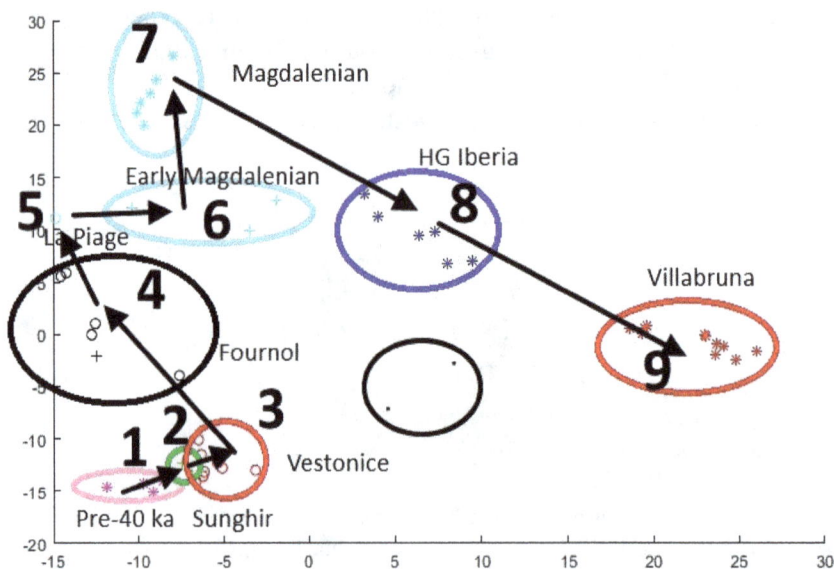

Fig. 5.4. MDS analysis of the f_3-data. The arrows and numbers represent the circular order of the ordered clusters obtained with a minimum contradiction approach (the two points in the middle have a high contradiction and were not classified).

ordered matrix and the admixture tree is a simple validation method that is quite useful for diagnosing some sensitivity issues.

Theoretical and experimental analysis (Recanati *et al.*, 2018; Otea-Garcia et Oteo, 2021; Peter, 2022) shows that the circular order of a high-dimensional character-state matrix compatible with a tree or phylogenetic structure is usually well visualized by projecting the matrix f_3, on a two-dimensional space. MDS and PCA are the most popular projection methods (Chapter 1). Figure 5.4 shows the order of the clusters, minimizing the contradiction plotted on a two-dimensional projection of the f_3 matrix. In the present example, the results of the multidimensional scaling reproduce the results quite well in terms of the ordering of the data. This result is not universal; otherwise, the problem would not be NP-hard. This example shows that one learns from observing the MDS projection or the PCA analysis.

5.3.2 *Classification of myth's motifs of the world*

The database by Berezkin (2012) is probably the largest database of myth and folktale motifs and contains the absence/presence of over 2200 motifs in almost a thousand different traditions. The database has specificities that make it ideal for areal studies. The motifs cover the whole world, contrary to other databases covering mainly Europe and Eurasia. The database selects primarily geographically widespread motifs often found on several continents. To understand how myth motifs are distributed and related across different cultures, we analyzed the Berezkin database.

The database can be represented as a very large matrix. The presence/absence of a motif is coded as one and zero. The database can be used in multiple ways to extract information on the world distribution of myths. An analysis with NeighborNet provided very interesting results. This analysis revealed two distinct groups of motifs with contrasting geographic distributions (Thuillard, 2018).

The motifs were classified and subsequently validated with a NeighborNet approach using a minimum contradiction approach completed by a validation module. The study (Thuillard *et al.*, 2018a) identified a first corpus of motifs with ten main clusters (Fig. 5.5). Each character is validated using the contradiction approach described in Chapter 3, setting a validation threshold. The procedure was repeated on the non-validated motifs, and to our surprise, a classification on this second corpus did fit quite well with another outer planar network topology (Thuillard *et al.*, 2018a). The two corpora cover the whole world and,

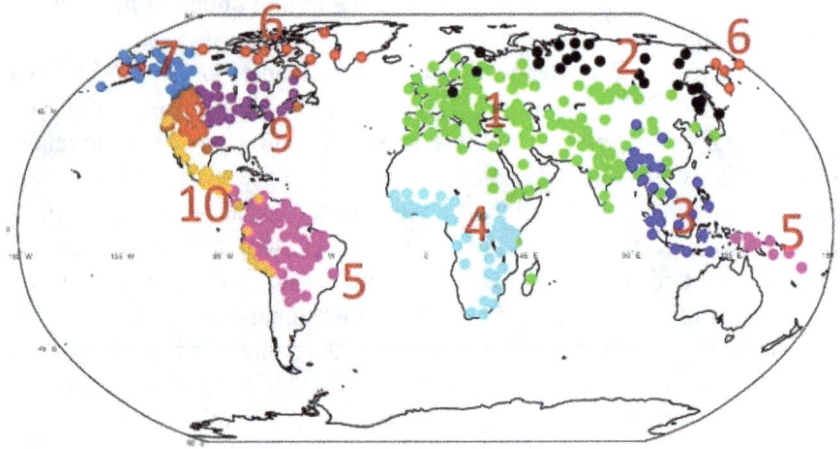

Fig. 5.5.   The first corpus of myth motifs has ten main regions.

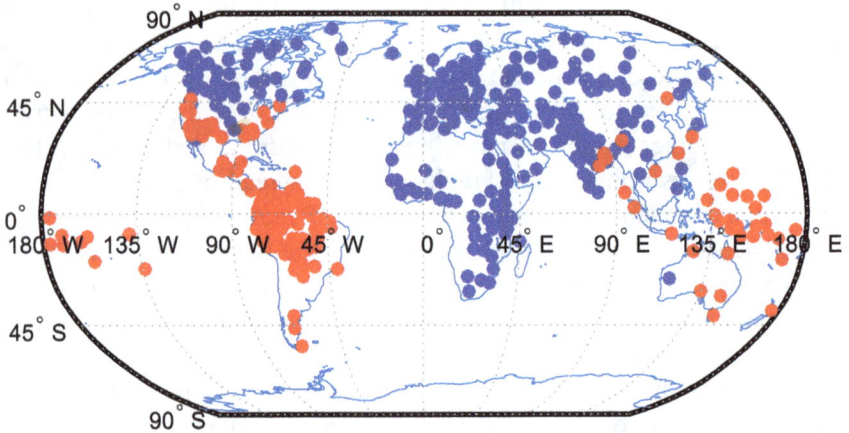

Fig. 5.6.   Most motifs can be classified into two corpora. The traditions represented by a blue dot have an over-representation of motifs associated with the first corpus, often linked to woman, Man, animal, or death. In contrast, the points in red have an over-representation of motifs associated with the second corpus, containing many motifs related to the sun and the Moon.

therefore, overlap geographically. The first corpus is over-represented in Eurasia and, to some extent, in Africa and North America (Fig. 5.6).

The second corpus is over-represented in South America and Papua New Guinea, around the Pacific coast. The result in Fig. 5.6 was reproduced independently recently using different techniques (Kim *et al.*, 2024).

About 15% of the motifs could not be classified. Further, the study identified that the motifs in the first group are predominantly associated with woman-related motifs, while the second group is strongly linked to sun and moon motifs.

The results' analysis is easier if the classification is done on the different traditions, as in Fig. 5.5, but also on the motifs. One observes that groups of motifs characterize the different clusters well. Clear deviations to a perfect phylogeny are observed (Cluster 11 is not valid, as the contradiction is large). Figure 5.7 suggests that many motifs have a large density within one of the ten regions. Some motifs expand over other regions.

We interpret the existence of the two corpora, detected with the phylogenetic network approach, as related to different diffusion paths into America. Figure 5.8 shows the distribution of motifs shared between reference traditions (red) and traditions in other parts of the world (blue). The left figure shows the regions with many shared motifs between the reference traditions in North America (red) and the rest of the world (blue). A tradition is shown if it has more than 60 shared motifs. The right part shows a similar analysis with the Amazonian area as a reference. Figure 5.8 suggests at least two diffusion paths for the motifs. Let us note that migrations (in both directions) did not stop after the physical

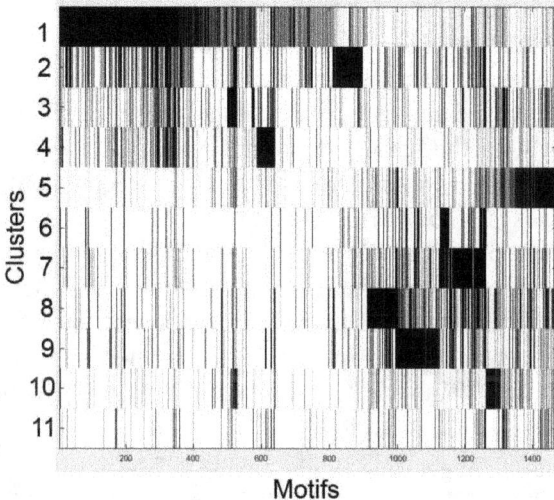

Fig. 5.7. Distribution of motifs in the ten main clusters (1–10) after clustering the traditions and the motifs in the first corpus. A motif in more than 15% of the traditions in a cluster is represented in black. The cluster numbers refer to Fig. 5.5.

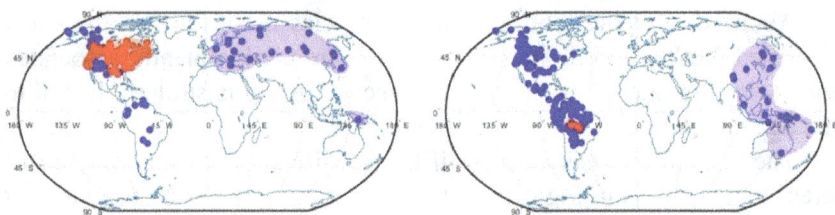

Fig. 5.8. Motifs distribution centered on North American traditions (left) and South American traditions (right). The red patch indicates the reference traditions, and the blue patches represent the traditions with the most shared motifs with the reference traditions.

connection between America and Eurasia ended. Therefore, it is difficult to date the events leading to the above results.

How are myth motifs transmitted between neighboring cultures? So far, we have looked at the big picture of how myth motifs are spread worldwide. Now, we will see how much these motifs are shared between neighboring cultures. The analysis uses an approach in the spirit of the work discussed above on the propagation of bread wheat (Zhao *et al.*, 2023).

After removing the average dependency on the geographic distance, one can map the motif distance between the neighbor traditions. More precisely, the distance between traditions is compared to the average distance between traditions as a function of geographic distance. The world is divided into small triangles, and the distances between the tips of the triangles are averaged and associated with a color code. Figure 5.9 identifies the areas with the strongest motif similarity between neighboring traditions. A better general picture is obtained by keeping only the highest similarity values (Fig. 5.9(b)). High similarities between regions are, for instance, observed between Scandinavia and England, in the subarctic areas in the Northern Hemisphere, or the Northwest of America. One also observes high similarity values in the Pacific between the different islands. Diffusion barriers in North America are associated with the mountains between the Great Plains and California and the region between Central America and the South of North America.

This fine-grained analysis complements our earlier findings on global motif distribution, providing a more nuanced understanding of how myths spread and evolve across cultures depending on geographic barriers. To what extent do other factors like distance or languages separate motifs?

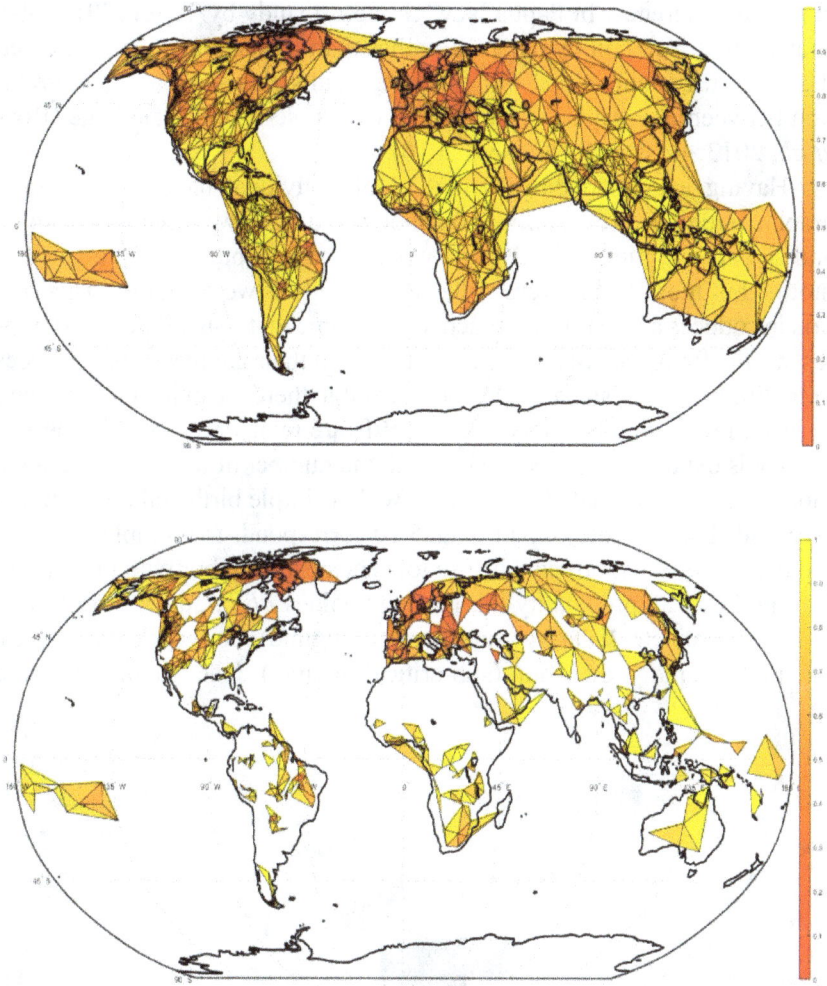

Fig. 5.9. (a) Effective diffusion surface. Reddish regions have more motifs in common than yellowish regions; (b) After thresholding and rescaling. The motif distance between two traditions equals the number of shared motifs divided by the total number of motifs observed in the two traditions.

Analyzing the databank, we find out that different traditions share, on average, 20% of the motifs at a distance of 500 km. This percentage drops to 10% at a distance of 2000 km. After repeating the experiment and limiting the statistics to the same language family, we only observed a marginal difference. In Papua New Guinea, a study by Terrel (2015) also showed that isolation by distance, rather than by language, has patterned their cultural relationships. For folklore, results also show a high correlation between geographic distance and, to a lesser extent, language (Ross *et al.*, 2013; Ross and Atkinson, 2016).

Having examined the distribution of individual motifs across geographical regions, we now focus on understanding how these motifs evolve and spread over time. Ideally, one would like to capture the dynamics of motifs, but most motifs were collected in the last two centuries. This section introduces a possible approach to overcome that issue. Figure 5.10 represents the frequency of motifs as a function of their number of occurrences in traditions in the database. One observes that there are many motifs found in just a few traditions. Only a few motifs are really widespread. The frequency is exponentially decreasing with the number of traditions sharing a motif. We have simulated the process with a simple birth and death model on a grid. The simulated data in Fig. 5.10 corresponds to a simple diffusion model, in which a motif can diffuse to its nearest neighbor with probability Pb and dies with probability Pd. For high values *Pb/Pd*, the motif diffuses over the complete 2D array (super-critical regime). For very small values, the motif rapidly dies out (sub-critical regime). The simulated curve

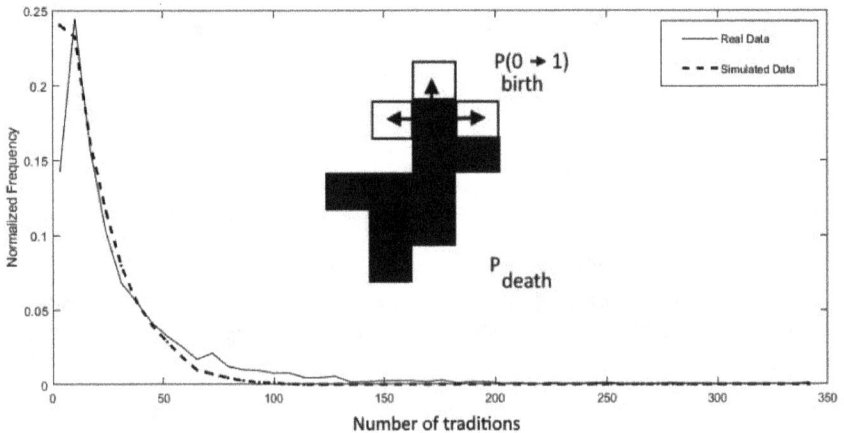

Fig. 5.10. Distribution of motifs among the different traditions. The simulated data correspond to a simple diffusion model with birth and death probabilities. The best fit is for parameters corresponding to the transition between the sub- and the super-critical regime.

corresponds to the region between the two regimes. If the motif were a virus, the distribution would correspond to a regime with no explosive growth nor a regime in which the virus dies out rapidly. It would be in the intermediary regime where the virus stays alive long but relatively localized. One observes deviations from this model for widespread motifs. More motifs are found on all continents than expected by the model. A possibility is that some motifs are in the super-critical regime. Examples are the motifs of the Man in the Moon or the motif in which the Sun and the Moon are married. The model predicts that the most widespread motifs must belong to the oldest ones on average.

This section shows that data analytics significantly complement classical comparative approaches by offering new hypotheses and a fresh view of Comparative Mythology. This approach requires multidisciplinary work combining mythological knowledge, areal studies, historical information, and multiple data analytics approaches.

5.3.3 *Application to the sky and starlores*

The previous sections have shown how useful phylogenetic networks can be to study myth distribution. In the next sections, we want to show the potential and limits of phylogenetic methods using concrete examples. Sky and starlore offer fascinating insights into how humans have interpreted the cosmos throughout history, and the examples are from that class of myths and folktales.

5.3.3.1 *Distribution of skylores*

The distribution of skylores in Fig. 5.11 (Sun, Moon, star, thunder, lightning, clouds, connections to the sky, descend/ascend to the sky, and many more) mirrors the distribution of the complete database. The main difference is that the classification has high contradiction values in Africa and subarctic regions in North America and could not be validated. The connection between North America and North Eurasia (Fig. 5.5) remains, but not between Amazonia and the Pacific regions.

5.3.3.2 *"Rabbit on the Moon"*

To illustrate the potential and limits of phylogenetic methods, let us examine the "Rabbit on the Moon" motif. On all continents, traditions have

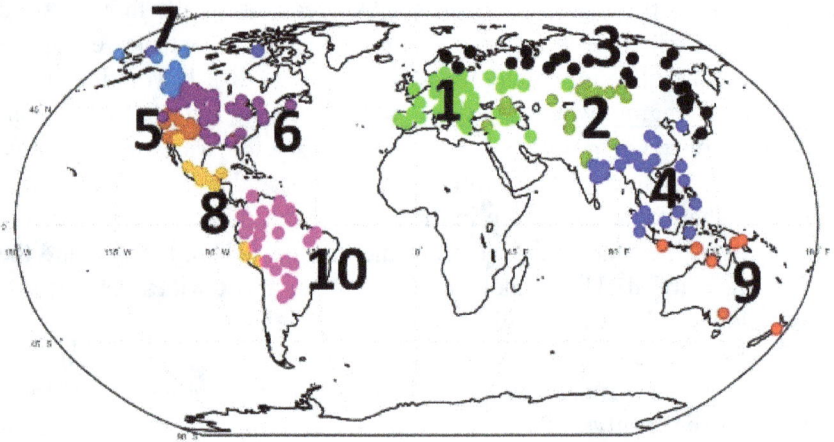

Fig. 5.11. Distribution of skylore motifs in the traditions recorded in the database by Berezkin.

stories about a man on the Moon. In many instances, a man is seen literally on the Moon. The dark patches on the lunar surface create the illusion. In large geographic areas, stories are also about a rabbit, a frog, or a water carrier, frequently a young woman carrying a pale.

We have analyzed the distribution of the "Rabbit on the Moon" and its transformation in a woman (or a boy) with a water pale as one moves from Africa to India, China, and around the subarctic back to Europe. The stories transform from place to place around the topic of death (Africa), self-sacrifice (India), and the elixir of life (China). The frog and the water motif is found in East Asia with a transition zone toward the "Water Carrier in the Moon" topic in Northern Eurasia, a motif also found in Shakespeare's Midsummer Night's Dream (Thuillard, 2021).

While the 'Rabbit on the Moon' geographical distribution reveals interesting patterns, its chronology is challenging. The time depth of the story is easily proven, but its chronology is very difficult to reconstruct, as most likely, the story predates considerably the first written mention of it. The 'frog/toad in the Moon' motif was recorded in writing in the Book of Changes (I Ching) over 2400 years ago. The "Water carrier on the Moon" theme is in old Norse literature in the Younger Edda. In Mesoamerica, the "Rabbit in the Moon" story is pre-Columbian.

Without the equivalent of a molecular clock, the exact dating of myths is impossible. Figure 5.12 shows the distribution of the different motifs.

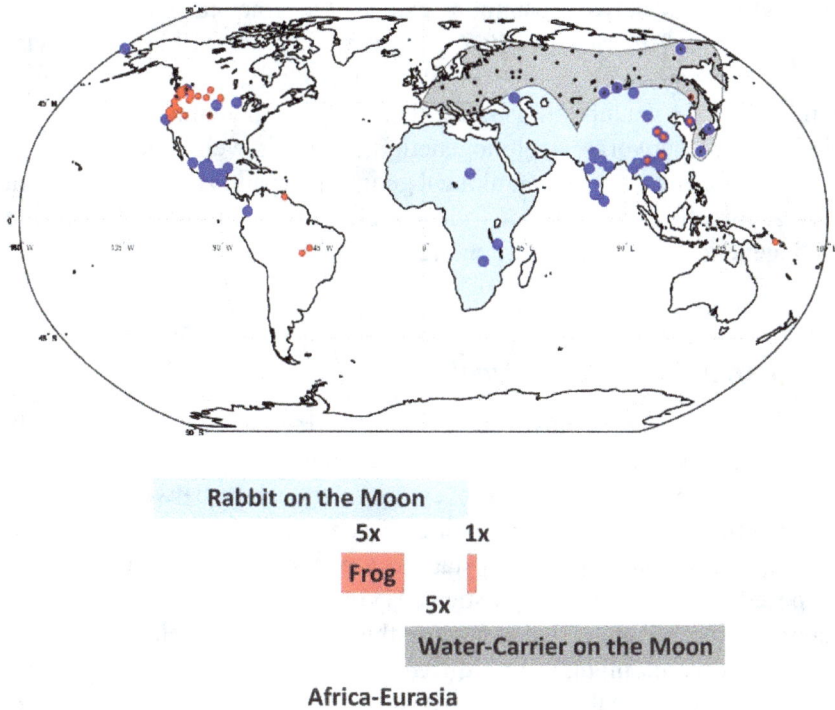

Fig. 5.12. (Top) The blue dots correspond to instances of the "Rabbit on the Moon" motif. The black dot and the gray area correspond to a "Water-carrier on the Moon" motif. Blue and gray dots overlapped in the gray area are instances with both motifs. The "Frog/ toad on the Moon" (red dots) is mostly encountered with the "Rabbit on the Moon" motif in Eurasia but not America. Bottom) The different motifs can be ordered, at a coarse geographic level, to almost fulfill the consecutive-ones condition in Africa and Eurasia, as sketched by the colored blocks, representing the presence of a motif. Modified from Thuillard (2021).

In Africa and Eurasia, the consecutive-ones condition is fulfilled quite well. The motifs often appear together at the geographic interface between two motif areas. In Africa and Eurasia, the data fit quite reasonably to a phylogenetic network with adjacent nodes corresponding to traditions from neighboring areas. A phylogenetic network description follows naturally from the diffusion and transformations of the different motifs.

The geographical consistency between geography and the motif distribution is disrupted when we consider the distribution of the motif in America. As Fig. 5.12 shows, the geographic proximity between nodes of the phylogenetic network cannot be extended to America. All three motifs are recorded in America, but the American traditions result in large deviations from a network or tree topology, breaking the consecutive-ones condition. This example illustrates, therefore, the potential and limits of phylogenetic networks. A phylogenetic network is well suited to motifs that diffused and transformed along a geographic path. If motifs diffuse in several directions and do not follow a clear path, large deviations to a phylogenetic network description are expected!

5.3.3.3 *Limitations of phylogenetic trees and networks in representing geographic diffusion*

Lévi-Strauss indirectly furnished support for tree representations in myth representation. The structuralist approach puts much weight on inversions, which primarily refer to reversing relationships between elements within a myth. In binary-state phylogenies, an inversion may be coded as the transformation from a zero state to a one state. Lévi-Strauss (2014) proposed another more sophisticated view of transformations in his canonical formula, intended to capture the underlying structure and transformations within myths across different cultures based on double inversions. The canonical formula can be represented as a quartet tree (Thuillard and Le Quellec, 2017) on ternary states (−1,0,1), further suggesting the applicability of tree structures in understanding myth transformations. Even if one assumes a tree topology, an important question is how well the tree fits the different traditions' geographic distribution.

Figure 5.13 shows a diffusion scenario (migration or propagation through interactions between neighbors). It illustrates how a phylogenetic tree can accurately represent geographical relationships when diffusion occurs primarily between neighbors but can become misleading when long-range migrations territorially overlap. The tree in Fig. 5.13 assumes two non-interacting diffusion waves, possibly at a different time. In a tree representation, if the branches do not cross when mapped geographically, neighboring nodes on the tree generally represent geographically close traditions. As long as a tree representation on the map preserves the topology with no edge crossing, adjacent nodes tend to be geographic neighbors. If people undergo long-range migrations, edges may intercept after geographic mapping. In that case, some adjacent taxa on the tree are geographically not neighbors. The above explanation may explain the data distribution in America (Fig. 5.12). It highlights the importance of

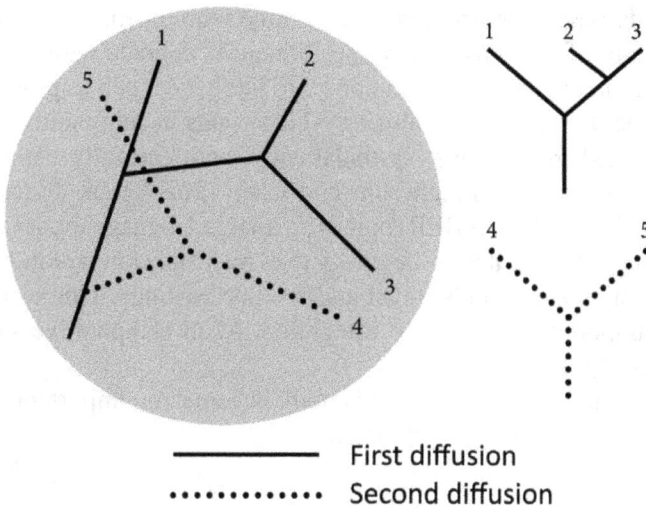

First diffusion

Second diffusion

Fig. 5.13. An example of a phylogenetic tree with a topology that matches geography. Edges cross, and the order of the taxa is not congruent with geography in the sense that some adjacent taxa are not nearest neighbors. Adapted from Thuillard *et al.* (2018b).

considering migration patterns and other factors that can influence the spread of myths, as they can significantly impact the interpretation of phylogenetic trees.

5.3.3.4 *LLM in comparative mythology?*

The last chapters analyze the world distribution of myths using a large matrix, indicating the presence or absence of a motif in traditions. Generating such a matrix is time-consuming and error-prone. In exploratory work, I sought to answer the question: Can LLMs (ChatGPT, Gemini AI, Claude) support this task, potentially accelerating research in comparative mythology? The answer is positive. LLMs demonstrate a capacity to extract motifs and generate a binary matrix coding the presence or absence of each motif.

Moreover, they can determine the latitude and longitude of different traditions, generate visualizations, and even utilize Python software to classify the matrix with a hierarchical clustering algorithm. LLMs can also assist in searching the literature for related stories. While promising, the results are still sub-optimal, with the LLMs sometimes misinterpreting complex narratives or struggling to identify subtle thematic connections. Important motifs are missed, and performances degrade with the number of versions analyzed. Future versions will likely be much improved. Even at this stage, LLMs can be valuable AI assistants in comparative mythology, provided researchers systematically and carefully validate the provided information. Tangherlini and Chen (2024) took a step in that direction. They combine BERTopic with a large language model tuned on 19th-century Danish literary language to extract and compare the topics in Hans Christian Andersen's travel and folktale writings. This work represents an important step toward integrating AI in comparative studies of folktales.

Despite its limitations, LLMs will become an important tool for research in Comparative Mythology.

5.3.4 *Other applications*

Outer planar networks have been applied outside of biology to many different problems. Let us give some significant examples. A study (Learmouth *et al.*, 2024) analyzes the cultural diversity of Pama-Nyungan speakers in Australia. The phylogenetic tree associated with traits related to initiation correlates well with the Pama-Nyungan phylogeny (Bouckaert *et al.*, 2018). Conversely, the data related to rock art or mortuary rites diverge from the language tree. In other words, it seems that initiation rites evolved with migrations within Australia in a parallel way to language. The data were tested for reticulation with NeighborNet. The results showed that the data better fit a network than a tree. Considering the prohibitive computational costs of Bayesian approaches to phylogenetic networks, a distance approach like NeighborNet is justified, especially since the circular order of the taxa on the network is, in that example, compatible with the one on the tree. Generally, comparing a tree obtained by a sophisticated approach to a distance-based phylogenetic network is usually insightful.

In linguistics, subgrouping language varieties with dialect continua is a major challenge. The different dialects do not fit well with a tree structure, and outer planar networks permit the representation of the complex structures between related dialects, as in the Lalo in China (Yang, 2010) and the Mixtecan language families (Auderset *et al.*, 2023).

A phylogenetic network may characterize cultural dependence and borrowing between different sites in archaeology. In a study on the fast spread of the Neolithic in the western Mediterranean (Rigaud *et al.*, 2018), data analysis demonstrates that characters defining pottery in different sites are well represented by an outer planar network, showing similarities and some local differentiations. On the other hand, personal ornaments were more or less identical on all sites.

Let us mention a study that probed the near past using phylogenetic networks. The study analyzed the evolution of skateboards. The history of skateboard development is well-known; therefore, the study could prove that a well-designed analysis allows for reconstructing skateboard evolution. The work is a good example of how specialists in a specific field can enhance the quality of a phylogenetic study, associating the main splits to particular economic and social contexts (Prentiss *et al.*, 2011, 2016).

Let us emphasize again an important aspect of phylogenetic networks that should be considered when interpreting results on non-biological data. A phylogenetic network may be interpreted as a phylogenetic tree representing a so-called vertical evolution with the addition of horizontal transfers in the form of admixtures, hybridization, or lateral transfers. A perfect phylogenetic network may also result from a different process in which characters appear at multiple locations and diffuse to their neighbors (see Fig. 5.2). The difficulty is that the two descriptions cannot be distinguished without knowing a chronology. Deviations from a tree topology leading to an outer planar network are often interpreted as (horizontal) transmission between groups, while (vertical) transmission is believed to be down some cultural lineages. It is a possible interpretation, but not the only one! Without further chronological information, a phylogenetic network may also result from the independent diffusion of characters out of multiple origin sites.

These diverse examples demonstrate the usefulness of outer planar networks in analyzing complex evolutionary patterns across a wide range of disciplines, provided careful consideration is given to interpreting the results.

5.4 Complement: On the Robustness of the F3-Statistics to Admixtures

Consider an ordered matrix representing the elements of the f_3-statistics with a reference taxon n fulfilling the Kalmanson inequalities. Let us show that a lateral transfer between consecutive taxa preserves the Kalmanson inequalities.

Proof.

Let us model an admixture with the equation:

$$p_i = \alpha p_{i-1} + (1-\alpha)p_{i+1} \text{ then, the condition}$$

$$f_3(n; i^*, j) \geq f_3(n; i^*, k) \ (i^* < j < k < n \text{ in a circular order) holds.}$$

Case: $i = i^*$: One has

$$f_3(n;i,j) - f_3(n;i,k) = \sum_{s=1}^{S} \left(p_{n,s} - p_{i,s}\right)\left(p_{n,s} - p_{j,s}\right)$$
$$- \left(p_{n,s} - p_{i,s}\right)\left(p_{n,s} - p_{k,s}\right)$$

and therefore

$$f_3(n;i,j) - f_3(n;i,k) = \sum_{s=1}^{S} \left(p_{n,s} - p_{i,s}\right)\left(p_{k,s} - p_{j,s}\right)$$

It follows that

$$f_3(n; i, j) - f_3(n; i, k)$$
$$= \sum_{s=1}^{S} \left(p_{n,s} - (\alpha p_{i-1,s} + (1-\alpha)\, p_{i+1,s})\right)\left(p_{k,s} - p_{j,s}\right)$$
$$= \alpha((f_3(n; i-1, j) - f_3(n; i-1, k))$$
$$+ (1-\alpha)f_3(n; i+1, j) - f_3(n; i+1, k)) \geq 0$$

Both terms are larger than zero for $0 < \alpha < 1$. The proof is almost identical for an admixture on n, j, k.

In studies, a negative f_3 value is generally regarded as indicating an admixture (Patterson *et al.*, 2012). Here, we use a different strategy to detect admixtures between non-consecutive taxa in a circular order.

References

Auderset, S., Greenhill, S. J., DiCanio, C. T., & Campbell, E. W. (2023). Subgrouping in a 'dialect continuum': A Bayesian phylogenetic analysis of the Mixtecan language family, *Journal of Language Evolution*, 8(1), pp. 33–63.

Berezkin, Y. E. (2012). Seven brothers and the cosmic hunt: European sky in the past, *Paar sammukest XXVI, Eesti Kirjandusmuuseumi aastaraamat* 2009, pp. 31–69.

Boc, A., Di Sciullo, A. M., & Makarenkov, V. (2010). Classification of the Indo-European languages using a phylogenetic network approach, *Classification as a Tool for Research: Proceedings of the 11th IFCS Biennial Conference and 33rd Annual Conference of the Gesellschaft für Klassifikation eV, Dresden* (Springer Berlin Heidelberg), pp. 647–655.

Boc, A., Diallo, A. B., & Makarenkov, V. (2012). T-REX: A web server for inferring, validating and visualizing phylogenetic trees and networks, *Nucleic Acids Research*, 40(W1), pp. W573–W579.

Bomfleur, B.G., Grimm, W., & McLoughlin, S. (2017). The fossil Osmundales (royal ferns)—A phylogenetic network analysis, revised taxonomy, and evolutionary classification of anatomically preserved trunks and rhizomes, *PeerJ* 5, e3433.

Bouckaert, R. R., Bowern, C., & Atkinson, Q. D. (2018). The origin and expansion of Pama–Nyungan languages across Australia, *Nature Ecology & Evolution*, 2(4), pp. 741–749.

Bryant, D., & Moulton, V. (2004). Neighbor-net: An agglomerative method for the construction of phylogenetic networks, *Molecular Biology and Evolution*, 21(2), pp. 255–265.

Coiro, M. (2024). Embracing uncertainty: The way forward in plant fossil phylogenetics, *American Journal of Botany*, 111(2), p. e16282.

d'Huy, J. (2013). A Cosmic Hunt in the Berber sky: A phylogenetic reconstruction of a Palaeolithic mythology, *Les Cahiers de l'AARS*, 15, pp. 93–106 (in French).

Gusfield, D. (2014) *ReCombinatorics: The Algorithmics of Ancestral Recombination Graphs and Explicit Phylogenetic Networks* (MIT Press).

Huson, D. H., & Bryant, D. (2006). Application of phylogenetic networks in evolutionary studies, *Molecular Biology and Evolution*, 23(2), pp. 254–267.

Huson, D. H., Rupp, R., & Scornavacca, C. (2010) *Phylogenetic Networks: Concepts, Algorithms and Applications* (Cambridge University Press).

Huson, D. H., & Scornavacca, C. (2012). Dendroscope 3: An interactive tool for rooted phylogenetic trees and networks, *Systematic Biology*, 61(6), pp. 1061–1067.

Kim, H., Hamilton, M. J., Jung, W. S., & Youn, H. (2024). Deeply nested structure of mythological traditions worldwide, arXiv preprint arXiv:2408.07300.

Learmouth, D., Layton, R. H., & Tehrani, J. J. (2024). The evolution of cultural diversity in Pama-Nyungan Australia, *Humanities and Social Science Communications*, 11, p. 945.

Legendre, P., & Makarenkov, V. (2002). Reconstruction of biogeographic and evolutionary networks using reticulograms, *Systematic Biology*, 51(2), pp. 199–216.

Lévi-Strauss, C. (2014). *Mythologiques* 1–4. Plon (in French).

List, J. M., Shijulal, N. S., Martin, W., & Geisler, H. (2014). Using phylogenetic networks to model Chinese dialect history, *Language Dynamics and Change*, 4(2), pp. 222–252.

Nakhleh, L., Ringe, D. A., & Warnow, T. (2005). Perfect phylogenetic networks: A new methodology for reconstructing the evolutionary history of natural languages, *Language*, 81(2), pp. 382–420.

Neureiter, N., Ranacher, P., Efrat-Kowalsky, N., Kaiping, G. A., Weibel, R., Widmer, P., & Bouckaert, R. R. (2022). Detecting contact in language trees: A Bayesian phylogenetic model with horizontal transfer, *Humanities and Social Sciences Communications*, 9(1), pp. 1–14.

Patterson, N., Moorjani, P., Luo, Y., Mallick, S., Rohland, N., Zhan, Y., ... & Reich, D. (2012). Ancient admixture in human history, *Genetics*, 192(3), pp. 1065–1093.

Petkova, D., Novembre, J., & Stephens, M. (2016). Visualizing spatial population structure with estimated effective migration surfaces, *Nature Genetics*, 48(1), pp. 94–100.

Pickrell, J. K., Pritchard, J. K. (2012). Inference of population splits and mixtures from genome-wide allele frequency data, *PLoS Genetics*, 8(11), p. e1002967.

Posth, C., Yu, H., Ghalichi, A., Rougier, H., Crevecoeur, I., Huang, Y., ... & Krause, J. (2023). Palaeogenomics of upper palaeolithic to neolithic European hunter-gatherers, *Nature*, 615(7950), pp. 117–126.

Prentiss, A. M., Skelton, R. R., Eldredge, N., & Quinn, C. (2011). Get rad! The evolution of the skateboard deck, *Evolution: Education and Outreach*, 4, pp. 379–389.

Prentiss, A. M., Walsh, M. J., Skelton, R. R., & Mattes, M. (2016). Mosaic evolution in cultural frameworks: Skateboard decks and projectile points. Straffon M.L. (ed.), *Cultural Phylogenetics: Concepts and Applications in Archaeology*, pp. 113–130.

Rigaud, S., Manen, C., & Garcia-Martinez de Lagran, I. (2018). Symbols in motion: Flexible cultural boundaries and the fast spread of the Neolithic in the western Mediterranean, *PLoS One*, 13(5), e0196488.

Ross, R. M., & Atkinson, Q. D. (2016). Folktale transmission in the Arctic provides evidence for high bandwidth social learning among hunter–gatherer groups, *Evolution and Human Behavior*, 37(1), pp. 47–53.

Ross, R. M., Greenhill, S. J., & Atkinson, Q. D. (2013). Population structure and cultural geography of a folktale in Europe, *Proceedings of the Royal Society B: Biological Sciences*, 280(1756), https://doi.org/10.1098/rspb.2012.3065.

Tangherlini, T. R., & Chen, R. (2024). Travels with BERT: Surfacing the intertex-tuality in Hans Christian Andersen's travel writing and fairy tales through the network lens of large language model-based topic modeling, *Orbis Litterarum*, https://doi.org/10.1111/oli.12458.

Terrell, J. E. (2015). Language and material culture on the Sepik Coast of Papua New Guinea: Using social network analysis to simulate, graph, identify, and Analyze social and cultural boundaries between communities, *Journal of Island Coast Archaeology*, 5, pp. 3–32.

Thuillard, M. (2009). Why phylogenetic trees are often quite robust against lateral transfers. In *Evolutionary Biology: Concept, Modeling, and Application* (pp. 269–283). Berlin, Heidelberg: Springer Berlin Heidelberg.

Thuillard, M. (2021). Analysis of the worldwide distribution of the 'Man or Animal in the Moon' motifs, *Folklore: Electronic Journal of Folklore*, (84), pp. 127–144.

Thuillard, M., & Le Quellec, J. L. (2017). A phylogenetic interpretation of the canonical formula of myths by Lévi-Strauss, *Cultural Anthropology and Ethnosemiotics*, 3(2), pp. 1–12.

Thuillard, M., Le Quellec, J. L., & d'Huy, J. (2018b). Computational approaches to myths analysis: Application to the cosmic hunt, *Nouvelle Mythologie Comparée/New Comparative Mythology*, 4, pp. 123–154.

Thuillard, M., Le Quellec, J. L., d'Huy, J., & Berezkin, Y. (2018). A large-scale study of world myths, *Trames: A Journal of the Humanities and Social Sciences*, 22(4), pp. A1–A44.

Warnow, T., Evans, S. N., & Nakhleh, L. (2024). Progress on constructing phylogenetic networks for languages, *The Method Works: Studies on Language Change in Honor of Don Ringe* (Cham: Springer International Publishing), pp. 45–62.

Wen, D., Yu, Y., Zhu, J., & Nakhleh, L. (2018). Inferring phylogenetic networks using PhyloNet, *Systematic Biology*, 67(4), pp. 735–740.

Yang, C. (2010). *Lalo Regional Varieties: Phylogeny, Dialectometry, and Sociolinguistics* (Doctoral dissertation, La Trobe University, Australia).

Zhao, X., Guo, Y., Kang, L., Yin, C., Bi, A., Xu, D., ... & Lu, F. (2023). Population genomics unravels the Holocene history of bread wheat and its relatives, *Nature Plants*, 9(3), pp. 403–419.

Part 4

Bayesian Methods in Artificial Intelligence

Chapter 6

Bayesian Approaches

6.1 Introduction

Bayesian analysis originates in the work of Bayes (1763), an English Reverand whose scientific work on probability was published post-humously. In the last decades, Bayesian analysis became a pillar of artificial intelligence. The basic equation behind Bayesian analysis is extremely simple, but its implications are far-reaching and not trivial. Application of Bayesian analysis requires manipulating probability distributions and sampling large data sets. Unlike classical or frequentist statistics, Bayesian statistics considers probability as a measure of belief or confidence in a hypothesis. The prior probability represents the initial belief or probability assigned to a hypothesis before observing any data.

A growing number of applications in sciences apply Bayesian approaches. In this book, the reader will find different applications of increasing complexity. The following chapters introduce Bayesian theory in a self-consistent manner. The various concepts and algorithms are presented directly using selected applications. Along the chapters, we will mention some software tools. Their use without much knowledge bears the risk of wrong or misinterpreted results due to algorithmic difficulties or misunderstanding the assumptions behind the different approaches. We believe that understanding the basic ideas and the pitfalls of the Bayesian approach is very helpful in avoiding such problems.

After a brief refresher on conditional probabilities and the Bayesian formula, the chapter introduces practical examples of dealing with uncertainty. Bayesian inference includes uncertainty at the core of its equations by assigning prior probability distributions to different hypotheses. The section about prior probabilities is illustrated with a complex problem: estimating the age at death of an ancient population from the state of the recovered bones in archaeologic sites. The chapter introduces multivariate Bayesian analysis with a regression problem to infer some aspects of the life of ancient humans or animals using isotope mixing models (Moore and Semmens, 2008; Parnell *et al.*, 2013). The chapter concludes with applications of Bayesian approaches in phylogenies.

Bayesian theory is becoming important in deep learning. Some algorithms were shown to perform similarly to Bayesian approaches without being implicitly coded (Kadkhodaie *et al.*, 2023). Dealing with uncertainty in deep learning is a central issue, and research is underway to purposely include a Bayesian framework in deep learning. The computing overhead is quite large; therefore, much work in that direction is necessary to optimize the algorithms (Papamarkou *et al.*, 2024), and discussing applications is premature.

6.1.1 *Conditional probabilities*

Let us start with simple but very important definitions. The conditional probability $P(A|B)$ is the probability of event A given event B. The conditional probability $P(A|B)$ can differ greatly from $P(A)$. The joint probability is the probability of two or more events: $P(A, B)$. There is a small but fundamental difference between conditional and joint probability. The joint probability is the probability of the simultaneous occurrence of multiple events. The conditional probability is the probability of one event, given that another event has occurred. The probability of a person in a population being 50 years old (event A) is much higher than the conditional probability $P(A|B)$ that a student (event B) is 50 years old. In this case, the joint probability $P(A, B)$ is the probability that a person is a student and 50 years old. The joint probability can be calculated as $P(A, B) = P(A|B) * P(B)$.

Given a joint probability $P(A, B, C,...)$, the marginal probability is computed in the discrete case by summing the probabilities on all or a limited number of events:

$$P(A) = \sum_{B,C...} P(A,B,C,...). \tag{6.1}$$

For continuous events, the sum is replaced by an integral:

$$P(A) = \int_{B,C...} P(A,B,C,...)dB\, dC\, ...$$ (6.2)

An essential concept in probability is the one of a random variable. A random variable describes the outcomes of a random process. A discrete random variable has values in a countable set, like the head and tail in a coin or the six faces of a dice. It could also be the number of students in a class. The probability distribution of a discrete random variable is often described using a probability mass function (PMF), which gives the probability of each possible value. Probabilities about rain in a simple weather report are often furnished in the form: "There is a 30% probability of rain tomorrow". More advanced reports furnish a probability density or PMF as forecast. The forecast in that form is more helpful to the reader. Figure 6.1 shows an example of a distribution.

A continuous random variable can take any value within a specified range. Its probability distribution is described using a probability density function (PDF). A continuous random variable may be the height of individuals or water temperature. The PDF is typically noted $f(x)$ for a single variable X. The function is non-negative on the range of definitions. If the definition range corresponds to the real numbers, then its integral sums up to one:

$$\int_{-\infty}^{\infty} f(x)dx = 1.$$ (6.3)

Sampling a random variable using a computer is an important process in Bayesian analysis. The sampling probability of X is proportional to the value of the PDF function. Values with higher PDF values are more likely

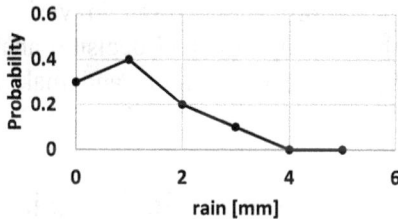

Fig. 6.1. Example of a PMF of the amount of rain during a day. The line relates the points for clarity.

to be sampled. For instance, uniform sampling is used for identical sampling of each value within the definition range. A Gaussian function is a continuous random variable from which data are sampled. A Gaussian function typically describes the probability of a physical random process like the measurement error or the wind speed distribution.

6.1.2 *The Bayesian formula*

Bayesian analysis is an old but essential formula for working with conditional probabilities. Bayes' formula states that

$$P(A|B) = P(B|A) * P(A))/P(B)), \qquad (6.4)$$

$P(A|B)$ is the conditional probability of event A given event B.
$P(B|A)$ is the conditional probability of event B given event A.
$P(A)$ is the probability of event A.
$P(B)$ is the probability of event B.

The Bayesian formula is often used in the context of having a model and data. Let us assume that A corresponds to a model explaining some data B. Bayes's formula becomes:

$$P(Model|Data) = P(Data|Model) * P(Model)/P(Data). \qquad (6.5)$$

The expression P(Data|Model) is defined as the likelihood. This equation tells us how to update our belief in a model after seeing some data. Bayesian inference is particularly interesting as it models uncertainty and integrates *a priori* knowledge. This prior knowledge might come from expert opinions, theoretical models, or previous experiments. The Bayesian approach uses the data to update prior knowledge and compute the posterior distribution. This process refines our initial beliefs, leading to more accurate probabilities for the model parameters.

Bayesian approaches are widely used in various fields, including machine learning, physics, biology, and decision analysis. They offer a coherent framework for updating beliefs and making decisions under uncertainty.

6.2 Introducing Constraints in Bayesian Inference

Simple constraints can be included in Bayesian inferences. Integrating time constraints is an interesting aspect of the Bayesian approach. As an

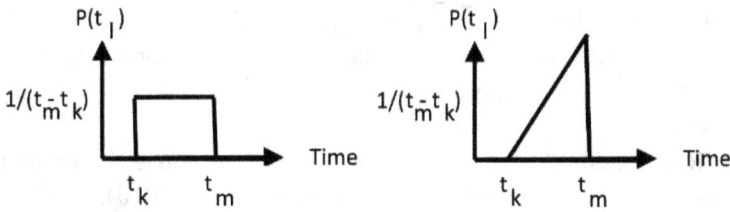

Fig. 6.2. The Bayesian framework allows the integration of constraints. Examples of time constraints are shown here, the first describing a uniform probability within some range while the second is biased toward large times.

illustration, let us assume that stratigraphic studies set relative constraints on the different samples used in the ^{14}C dating.

Let us consider the case in which the data comes from different time horizons. The stratigraphic information can be integrated into the Bayesian analysis. A uniform prior on the relative constraints is of the form:

$$p(t_l) = \begin{cases} \frac{1}{(t_m - t_l)} & t_k < t_l < t_m \\ 0 & \text{otherwise} \end{cases}. \tag{6.6}$$

The prior is integrated multiplicatively in the above equations.

Figure 6.2 shows left a uniform prior and right a prior with $t_k < t_l < t_m$. In the second instance, the prior reflects the information that the t_l is believed to be closer to the upper than the lower bound. These two examples show the simplicity and flexibility of introducing prior knowledge through constraints in a Bayesian analysis. Software packages like OxCal (Ramsey, 2009) integrate the possibilities to add constraints.

6.3 Priors in Bayesian Analyses

Paleodemography studies the demographic characteristics of past human populations. It aims to reconstruct ancient populations' size, structure, and dynamics using skeletal and archaeological data. Skeletons are often the only source of information that can be used to determine age at death in ancient populations. Bones and teeth indicate a deceased person's living conditions.

Determining the age of death from the conditions of the bones is very difficult, especially for extinct populations. Age and bone conditions are not well correlated, and large variations are observed from individuum to individuum.

Different methods determine the age distribution based on maximum likelihood or a Bayesian approach (Caussinus *et al.*, 2010). Studies start from a reference table giving bone or teeth conditions as a function of age at death. Table 6.1 shows an example.

The conditional probability distribution $P(\text{age} \mid \text{bone state} = S)$ derived from that table gives the age distribution for an individual with bone state S. When bones of many specimens are recovered in an excavation, more accurate results can be obtained than simply averaging the age distributions. Caussinus *et al.* (2010) use two prior distributions: one on the age distribution $g(p)$, where p represents the age probabilities, and another on $G(P)$, the conditional probabilities of observing specific bone states given an individual's age. The prior probabilities reflect the researcher's initial beliefs based on archaeological context or data from similar populations. The likelihood assumes that the observed bone state frequencies m_i follow a multinomial distribution with parameters π_i, representing the probability of observing bone state i in the population with the probabilities obtained from the table. The distribution of bone states recovered during excavation enters the multinomial's exponents. The joint density is given by

Table 6.1.   The Bayesian approach starts from a distribution of bone conditions and age at death.

Age	0–5	5–10	...	>50	Total by stage	Observed population
Bone conditions						
C_1	$n_{1,1}$	$n_{1,2}$	$n_{1,1}$	$n_{1,11}$	$n_{1,.}$	m_1
C_2	$n_{2,1}$	$n_{2,2}$	$n_{2,j}$	$n_{2,11}$	$n_{2,.}$	m_2
...			$n_{i,j}$		$n_{i,.}$	...
C_k	$n_{k,1}$	$n_{k,2}$	$n_{k,j}$	$n_{k,11}$	$n_{k,.}$	m_k
						Total Pop.:m
Total by age	$n_{.,1}$	$n_{.,2}$	$n_{.,j}$	$n_{.,11}$	n	

Source: Adapted from Caussinus *et al.* (2010).

$$f(M,P,p) = g(p)G(p)\frac{m!}{\prod_i m_i!}\pi_i m_i,\qquad(6.7)$$

with $m = \sum_i m_i$.

The last terms represent the multinomial distribution, which is a generalization of the binomial distribution to more than two possible outcomes (like heads/tails). The binomial distribution describes the probability of obtaining n successes in independent trials. The posterior probabilities can be computed from the joint density.

The multinomial distribution is appropriate for categorical count data, like here, with the different states of the bones that can be described linguistically (high wear, no wear). Different problems require different distributions. Continuous models like those in the next section often use a Gaussian distribution.

The example demonstrates the power and flexibility of Bayesian methods for making inferences from limited and imperfect data. Bayesian approaches are core methods in artificial intelligence. The main difficulty of the Bayesian approach is defining the prior probabilities, which requires much know-how. Bayesian methods permit improving domain-specific information by combining prior knowledge with experimental data to update the information. One of the strengths of Bayesian approaches is that researchers may refine these prior probabilities as they further learn about the specific archaeological context. The method is not without difficulties. A recent review explains some difficulties with the Bayesian approach in paleodemography (Boldsen *et al.*, 2022). In the sample population, age at death may not be easy to determine if the bones are heavily damaged or not well preserved. There may be sample bias due to the differential conservation of bones of children compared to adults or due to burying practices. The small bones of children are more likely to be lost than the bigger bones of adults, leading to infant underrepresentation in archaeological samples. Łukasik *et al.* (2021) use a Bayesian approach to estimate the demography of Scythian populations. Based on 312 individuals, they combine the information on bones and teeth growth (pre-adults) to measurements of tooth crown wear, morphology of the pubic symphysis, and cranial suture closure to describe the age at death for different skeletal stages. The study implements the Bayesian procedure (Caussinus *et al.*, 2010), adding correction factors for the age distribution of children. Validating Bayesian results is difficult, and there are no real universal methods. Algorithmically, testing the method's sensitivity to the prior is one of the main approaches to validating the reproducibility of the results. Łukasik *et al.* (2021) found out that results are quite sensitive to the assumptions on population growth. Despite this difficulty, they could extract a robust result from the data, namely a peak in mortality in early adulthood despite indicators of good health.

A new type of clock based on epigenetics allows the dating of the age of death of humans or animals. This approach, known as an epigenetic clock, utilizes DNA methylation to predict age at death (Horvath, 2013; Pederson *et al.*, 2013). After its discovery, the approach was rapidly applied to ancient human and animal DNA. Applications to ancient horses use bones and teeth (Liu *et al.*, 2023). Algorithmically, the difficulty is that modern and ancient horses have different measured methylation levels. This discrepancy arises from several factors, primarily DNA degradation and the influence of varying environmental factors over time. A calibration curve is necessary. Liu *et al.* (2023) use a random forest to compute the "ticking" rate of the clock. The random forest approach (see Chapter 1) is appropriate here as the bones and teeth furnish different methylation and ages. The output age estimate is obtained through a voting process (for continuous values, voting corresponds to a simple average or an average limited to the peak of the distribution). Comparing the results for horses of known ages showed an overestimation of the age of young horses. One understands that the method faces the same challenges as determining ages using the bones and teeth wear, and Bayesian methods are a good approach to dealing with uncertainty (Xu *et al.*, 2019).

A sex differential and a different methylation level in castrated horses introduce another level of complexity. An SVM algorithm (see Chapter 1) is typically used to separate the different outcomes (Liu *et al.*, 2023). Determining age, sex, and possible castrations opens new possibilities for understanding the early stages of domestication and the relations between herders and animals over the years.

In summary, a problem that first appears simple is an active and highly complex research topic with many caveats due to statistical biases in the data.

As a concluding remark, let us note that in Bayesian studies, the availability of reliable prior knowledge is of central importance. Bayesian approaches are meant to integrate the best prior knowledge to analyze data and are regarded as superior to maximum likelihood. However, the quality and accuracy of prior information significantly influence the reliability and validity of Bayesian inferences. Data analytics has no free lunch, so wrong priors generally lead to worse results than the maximum likelihood. Therefore, careful selection and validation of prior knowledge are crucial steps in any Bayesian analysis.

6.4 Multivariate Bayesian Analysis

Multivariate Bayesian analysis deals with multiple variables and makes inferences about the relationships between these variables, considering uncertainty. Bayesian linear regression models are the simplest form of such models. They describe an output as a weighted sum of input variables and determine the best parameters of the distributions associated with each variable with a Bayesian approach.

For example, Stable Isotope Mixing Models quantify the proportional contributions of various sources to a mixture. An important application is analyzing organisms' diets based on isotope traces. These models consider the metabolism of multiple tissues and organs. Moore and Semmens (2008) developed a Bayesian mixing model to estimate the different contributions to the nutrition of salmons. The model can also be applied to humans. The isotopic signature of a fish is a mixture of different isotopes k resulting from the proportion of the sources of food with its isotope fraction and an offset describing isotope enrichment due to the different chemical and physical properties of the element's different isotopes. One has

$$H_k = \sum_i \alpha_i (f_i + T_k). \tag{6.8}$$

The equations can be solved using the Bayesian equations using prior information furnished by expert knowledge. One obtains a predicted isotope signature from an initial belief about the different food contributions (prior). The likelihood of the observed data can be computed using a Gaussian function, which shows the difference between the predicted and the observed isotope signature. Together with the prior probabilities, one calculates the posterior probability. The combination of source contributions with the highest posterior probability represents the most likely scenario based on the data and prior information. The FRUITS (Fernandes *et al.*, 2014) software permits complex computations of the mixing parameters based on an extensive databank.

A model improvement consists of replacing the isotope fraction in the food source i by a term $(C_{i,j}I_{i,j,k})$ describing the kth isotope content in the jth food fraction (protein, carbohydrates, lipids, single amino acids) of the ith food group (plants, animals, fishes). Further, a term is introduced to consider the routing of the food fraction for the kth isotope.

$$H_k = \sum_j W_{j,k} \sum_i \alpha_i (C_{i,j}I_{i,j,k} + T_k). \tag{6.9}$$

The more refinements one introduces to a model, the more data are necessary on the differential incorporation of isotopes depending on the food chain (or food routing) and different tissues!

Multivariate Bayesian analysis can be computationally intensive, especially for large datasets or complex models. Markov Chain Monte Carlo (MCMC), particularly the Metropolis-Hastings algorithm described in the next section, is commonly used for posterior inference, but it can be slow and require careful tuning. The first applications of Bayesian analysis in archaeology are already quite old (Buck *et al.*, 1996; see Otárola-Castillo *et al.*, 2023). The use of Bayesian techniques is far from universal, with many studies relying on classical regressions, but the number of studies using Bayesian approaches steadily grows. Isotope analysis is a good application for understanding some difficulties in developing a good model, especially in the Bayesian context, in which prior knowledge is required. The information that may be gained is so interesting that Bayesian modeling is worth the effort!

Radioactive elements, such as strontium, provide valuable insights into a living organism's life. Strontium, with a chemistry related to calcium, is absorbed in tissues depending on the location. The strontium content varies significantly with rock age and chemical composition. The past living conditions of a person can be deduced from isotope measurement in bones and teeth. Many organic tissues grow continuously and record time series of $^{87}Sr/^{86}Sr$ variations that can be used to reconstruct movement or migration patterns of individuals or entire populations. Some quantity is transferred to the animal and food eaten by humans or the water they drink. Strontium is, for instance, integrated into tooth enamel during early childhood and becomes sealed there. After its death, strontium is not absorbed anymore and decays similarly to ^{14}C. The research follows several tracks, understanding the transfer of strontium to soils or water and then to the eaten food. The strontium intake depends on diet, and it is not always simple to separate the location from the diet effect. The similarity between strontium and calcium results in high levels in milk and dairy products. Seafood, especially "filter feeders" like shellfish, are also sources of strontium (Lahtinen *et al.*, 2021).

In recent years, the accumulation of data and the availability of software tools have simplified Bayesian techniques. For example, Gigante *et al.* (2023) determined the geographic origin and mobility of archaeological and modern fauna collected from Phoenician-Punic sites using a Bayesian provenance assignment workflow, including data represented by isoscapes, which are maps representing the isotope levels in some regions.

Different isotopes, such as carbon, nitrogen, oxygen, and hydrogen, are also used as markers to reconstruct the type of food eaten by animals or humans. Intakes depend on the kind of food, the food chain, and the tissues. Stable isotopes change relatively predictably during biological processes, and mass-balance mixing models can quantitatively assess the relative contribution of different sources to the mixture of interest (Makarewicz and Sealy, 2015).

The most critical aspects of Bayesian multivariate modeling are good data, correct priors, and a reliable model that captures the most important variables.

6.5 Metropolis-Hastings Algorithm

A Bayesian approach aims to furnish a posterior probability of the model given the data.

$$P(Model|Data) = P(Data|Model) * P(Model) / P(Data) \quad (6.10)$$

One assumes here that the prior probabilities are given, and the likelihood can be computed. The denominator in Eq. (6.10) is generally very difficult to calculate and often intractable. The Metropolis-Hastings algorithm (Metropolis *et al.*, 1953; Hastings, 1970) implements a clever approach to computing the posterior distribution without calculating the denominator using a Markov Chain Monte Carlo method.

The idea of the Markov Chain Monte Carlo (MCMC) is to sample new parameters dependent on the results of the current parameters. MCMC is so designed that the process explores the space of parameters to spend more time in the posterior's high-probability regions so that the distribution of samples after many steps corresponds to the posterior distribution in the Bayes equation.

The Metropolis-Hastings algorithm
Initialization: Choose an arbitrary starting state.
    At each iteration:

-   Generate a candidate state, x_{i+1}, sampled from a proposal distribution $g(x_{i+1}|x_i)$, for instance, a symmetric Gaussian centered around the current state.
-   Compute the posterior probability ratio $\beta = \frac{p(x_{i+1})}{p(x_i)}$ from the likelihood and prior probability distribution.
-   The probability of keeping the new sample is set to $\alpha = \min(1, \beta)$. Draw a value u among a uniform distribution in [0, 1]. Keep the new sample if $u > \alpha$ or else keep the old sample.
-   Stop the iteration when a stopping criterium is fulfilled. The posterior probability distribution is estimated from the sample distribution.

The acceptance criterion is designed so that the sample distribution tends toward the posterior distribution, provided there is no cycle, and the search can reach any state (Hastings, 1970). With this approach, the computation of the denominator in Eq. (6.10) is unnecessary, as learning depends on the ratio of the posterior distribution. Generally, the first values (burn-in) are removed from the statistics.

Using a prior in a Bayesian search is a great advantage compared to a maximum likelihood approach. The prior focuses the search on the interesting regions of the parameter space and reduces the necessary computing time to reach an optimal solution. If the prior is badly chosen, the prior will hinder the search efficiency. The search will focus on the wrong regions; on average, the search will be worse than without any prior, a direct consequence of the No Free Lunch Theorem (Wolpert and Macready, 1995). Also, if the prior is zero on the best parameters, then the Metropolis-Hastings algorithm does not explore this part of the space.

The choice of the variance of the Gaussian distribution used for sampling is also of great importance. Taking a too-small value may result in the search being stuck in local minima, and making the Gaussian too broad may lead to missing sharp maxima. Several approaches are used to assess the results. Repeating the search several times with different parameters and comparing the results is one central approach to determining if the results are reproducible and converge toward a stationary distribution.

Software packages like BEAST (Bouckaert *et al.*, 2014b) and RevBayes (Ronquist and Huelsenbeck, 2003) are powerful tools for implementing Bayesian approaches. They implement the Metropolis-Hastings algorithms and other strategies like Gibbs sampling, an alternative approach in which a single parameter is modified at each step (Resnik and Hardisty, 2010).

6.6 Bayesian Approach to Phylogeny

Bayesian approaches represent the state-of-the-art in phylogeny studies. In a Bayesian analysis, a prior describes tree topology, branch length, or molecular clock to model varying branch evolution rates. The computing time increases with the size of the parameter space. For that reason, Bayesian approaches are often limited in the number of parameters. The maximum number of taxa is generally much smaller than in other approaches, such as distance-based phylogenies.

6.6.1 *Application to languages*

In many studies, the phylogenetic analysis of languages relies on a list of cognates. There are widespread words with a common origin, like foot in English and Fuss in German. Cognates descend from a common ancestor word in a proto-language. Historical linguists define lists of cognates based on phonology and morphology using the classical methods of comparative linguistics (Greenhill *et al.*, 2020). The cognate list is the basis for phylogenetic modeling. Advanced models combine the covarion model (Section 4.7.1) with a birth and death model (Section 4.11) for modeling the tree (Heggarty *et al.*, 2023). The models are typically applied to model cognate transformations.

Greenhill *et al.* (2020) point to a strong concordance between classical comparative methods and phylogenetic analysis for many languages. This concordance was used in studies on the different hypotheses on the origin of Austronesian languages and their propagation from Taiwan (Greenhill *et al.*, 2010).

The origin of the Indo-European (IE) language family has been controversial for over a century (For a summary, Heggarty *et al.*, 2023, supplementary data). The question of origin has two main facets: the geographic origin of the IE language and the language itself. A break-through came as information from ancient DNA showed the existence of a massive migration from the steppe region into Western Europe with ancient DNA related to modern populations speaking Indo-European languages. The phylogeny of IE languages has produced conflicting results, partly due to the varying data quality. Heggarty *et al.* (2023) have applied the fossilized birth-death model described in Section 4.7.2 to find and date the common ancestor to IE languages. In the study, the dead languages are equivalent to fossils, and the model includes them consistently in the Bayesian approach.

The study differs from older studies in that language specialists coordinated their efforts to develop an improved IE core vocabulary database with a list of cognates. The results of that study are inconsistent with a pure steppe hypothesis, at least for the linguistic aspect, as the study suggests an emergence of IE languages around 8,000 years before the present.

The above approach is not suited to all languages. Many Southeast Asian and South American languages have words formed by compounding (Wu and List, 2023), which is the creation of a new word from two or more words. Compounding makes cognate identification quite complex.

6.6.2 *Total-evidence analysis*

Many recent studies emphasize the necessity of including information on evolution in the form of fossils in paleontology studies or ancient cognates in historical linguistics to date the nodes in phylogenetic trees. Dating is also sometimes possible using pottery and artifacts. Michael *et al.* (2022) apply the latter approach to Arakwan languages based on the assumption that the geographical radiation of Arakwan languages was associated with the appearance of novel material forms.

The fossilized birth-death model, illustrated in Fig. 4.25, integrates molecular data with morphological data on fossils and extant species in a single Bayesian framework (Barido-Sottani *et al.*, 2023). The model elegantly solves the dating issue caused by irregular speciation rates by including fossils with some dating constraints. The addition of fossils helps calibrate the speciation rate (Wright *et al.*, 2022).

The model consists of three blocks describing molecular evolution, morphological evolution, and the tree model. Molecular evolution typically uses a generalized-time reversible (GTR) transition rate matrix, the most general time-reversible matrix for DNA evolution. The model is completed by Gamma distribution (GTR+Γ) to account for the rate heterogeneity among sites (Felsenstein, 2004).

The evolution of morphological characters implements the Mk model (Lewis, 2001) and the fossilized birth and death model for tree modeling. The likelihood has three components that are multiplied to get the likelihood:

Molecular evolution:	$P(DNA\ Data \mid GTR - \Gamma, molecular\ clock, tree)$
Molecular evolution:	$P(Morphology \mid Mk, clock, tree)$
Tree:	$P(fossils, tree \mid FBD\ parameters)$

The posterior probability can be computed given the different priors and the above likelihoods.

Ronquist *et al.* (2012) introduced the "total-evidence analysis" in Bayesian studies, applying it to Hymenopters, the largest order of insects, including ants and bees. By including fossils, the study suggests that Hymenopters started diversifying into major extant lineages much earlier than previously thought, well before the Triassic. The problem of dating the apparition of the hymenopters is still much debated, and recent studies focus on dating some specific innovations like a form of parasitism widespread in Hymenopters in which the parasite cannot complete its life cycle without a host. Differences between total-evidence analysis and maximum likelihood studies on large databases differ by (only) about 20–30 Mio years (Blaimer *et al.*, 2023).

It is still very difficult to assess the merits of the total-evidence approaches. Compared to models with only topological constraints on the evolution tree, total-evidence approaches require acquiring and treating morphological data on fossils and their approximate dating. While a total-evidence analysis with a Bayesian approach seems superior to a maximum likelihood approach on molecular data, the uncertainty on dating early fossils and their classification is a potential source of error.

Results depend on the details of the models, which become increasingly complex and difficult to compare objectively. In that context, simulation studies on the FBD model play a central role in validating models. Simulations show that accuracy and precision strongly depend on the quality of the fossil data (Warnock *et al.*, 2017; Barido-Sottani *et al.*, 2023). In particular, simulations showed better results with large realistic time intervals for fossil datings than using inexact "point" dating. This result illustrates well the Bayesian approach's usefulness in integrating uncertainty (Warnock *et al.*, 2017).

The prior root age is also central to the quality of the results (Luo *et al.*, 2023). Beaulieu and O'Meara (2023) argue that some improvement in the model could be achieved by better integrating fossils of extant species in the computation of the sampling over time. All these results remind us of the importance of fossils and paleontological studies for tree calibration and the difficult issue of sampling in statistical approaches. Sometimes, a new fossil changes the results of the FBD model. A new fossil suggested, for instance, that ray fishes, or at least some of their characteristic features, were already present before the Devonian mass extinction (Giles *et al.*, 2023), leading to a revision of the tree proposal.

The FBD process is implemented in the software package BEAST (Gavryushkina and Zhang, 2020; Bouckaert *et al.*, 2014b).

6.6.3 *Phylogeography*

Bayesian analysis allows for the seamless integration of different aspects related to various characters like DNA and morphology. The model may also include parameters related to geography. Phylogeography combines geography and DNA to understand species' spatial and temporal distribution. For instance, if one is looking for the geographic origin of some bird species, the model may include a diffusion parameter with different transition rates for flying and flightless birds depending on the distance. Another approach (Lemey *et al.*, 2010) consists of modeling dispersal as a random walk, considering the geographic position of the extant taxa and the tree topology. In linguistic studies, one tries to determine the likely geographic origin of a language (Tao *et al.*, 2023) or the subsistence mode of the population (Greenhill *et al.*, 2023) at the tree root.

Softwares like BioGeoBEARS (Matzke, 2013) or BEAST make the application of Bayesian methods more accessible. Software tools include different speciation modes, such as founder-event speciation, in which a small group moves to a different area isolated from its original population, leading to speciation. The transition matrix is often limited to a few regions to ensure computational tractability.

6.6.4 *Results presentation*

The results of a Bayesian study are often difficult to interpret and, therefore, require an appropriate presentation. A common approach consists of

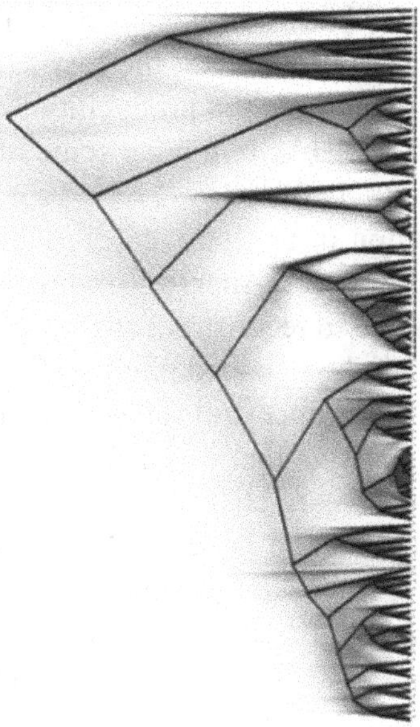

Fig. 6.3. Consensus tree overlapping the DensiTree.
Source: Adapted from Tao (2023). CC-NY.

reducing the distribution to a single tree, facilitating the interpretation of the results. A tree with the highest posterior probability in the sampled distribution is called the Maximum a Posteriori (MAP) tree. Given the data and priors, it is the "best" tree in a Bayesian framework. Another approach represents the branching patterns supported in over 50% of the sampled trees. Much information on the uncertainty and alternative solutions is lost in that representation. A DensiTree representation (Fig. 6.3) may be quite helpful in understanding the results. DensiTree draws all trees in the set simultaneously using semi-transparent lines (Bouckaert and Heled, 2014a).

6.6.5 *Validation of Bayesian models*

Bayesian phylogenies are still an intensive research field. The motivation lies in regular discrepancies observed between some Bayesian approaches and results obtained through classical methods, for instance, on the Indo-European language tree. Let us explain some of the current research directions.

Louca and Pennell (2020) state that dated trees are consistent with myriad diversification histories. Heggarty *et al.* (2023) address that issue by constraining the birth and death rates to be almost equal. Recently, Truman *et al.* (2024) proved that the fossilized birth-death model with a positive sampling rate (i.e., the probability that a lineage will be sampled is non-zero) is identifiable, removing some of the concerns introduced by Louca and Pennell (2020).

The computation of ancestor states at the root is even more difficult than determining the root. Ancestral states may be absent in extant taxa, making it impossible to reconstruct ancestral states, or a character may be non-congruent with the obtained phylogeny, which may significantly bias the computation of the ancestral state (Duchêne *et al.*, 2014). Gascuel and Steel (2020) have shown that it is usually impossible to estimate the root state and the rate of state change accurately from the extant taxa alone. If the rate of state change is low, the reconstruction of the root state is easy, but there are too few changes to estimate the rate of change. Inversely, evaluating the root state is difficult for large rates of state changes. In that sense, integrating dated fossils helps estimate the root states and the rate of state change.

Another issue is sampling. With simulations, Beaulieu and O'Meara (2023) showed that under-sampling old fossils on an extant line may profoundly influence the speciation and extinction rate, biasing the model towards a larger extinction rate. Despite furnishing biased values, the results may show high confidence values on the tree topology and parameters. Didier and Laurin (2024) agree on the dangers of integrating only a limited, biased portion of the fossil record. Still, they emphasize the fossil record's usefulness in studying temporal variations in speciation and extinction rates.

Bayesian approaches estimate parameters in distributions, assuming that regularities in some parts of the evolution tree may be used to estimate parameters or distributions on different tree sections. Difficulties arise when deviations in the assumptions behind a tree's evolution affect the parameters on some branches differently (e.g., rates of evolution, molecular clock). The above discussion shows that the validity of the models depends on what may look like details. Despite all the above discussions, Fossilized Birth and Date models are state-of-the-art methods that truly expand the questions that artificial intelligence can reliably address about our past. The Bayesian approach pushes the limits on what can be reconstructed from the past, making the result the best guess that is obtainable in the state of knowledge. The technique requires a lot of computing time and is, therefore, still limited in the number of taxa and samples. The problem is that results sometimes seem very reliable and reproducible, though they do not seem compatible with classical approaches. Could it be that, in some instances, one pushes the Bayesian methods beyond what is feasible given the data?

References

Barido-Sottani, J., Pohle, A., De Baets, K., Murdock, D., & Warnock, R. C. (2023). Putting the F into FBD analysis: Tree constraints or morphological data? *Palaeontology*, 66(6), p. e12679.

Bayes, T. (1763). An essay towards solving a problem in the doctrine of chances. By the late Rev. Mr. Bayes, FRS communicated by Mr. Price, in a letter to John Canton, AMFR S, *Philosophical Transactions of the Royal Society of London*, 53, pp. 370–418.

Beaulieu, J. M., & O'Meara, B. C. (2023). Fossils do not substantially improve, and may even harm, estimates of diversification rate heterogeneity, *Systematic Biology*, 72(1), pp. 50–61.

Blaimer, B. B., Santos, B. F., Cruaud, A., Gates, M. W., Kula, R. R., Mikó, I., ... & Buffington, M. L. (2023). Key innovations and the diversification of Hymenoptera, *Nature Communications*, 14(1), p. 1212.

Boldsen, J. L., Milner, G. R., & Ousley, S. D. (2022). Paleodemography: From archaeology and skeletal age estimation to life in the past, *American Journal of Biological Anthropology*, 178, pp. 115–150.

Bouckaert, R. R., & Heled, J. (2014a). DensiTree 2: Seeing trees through the forest, BioRxiv, 012401.

Bouckaert, R., Heled, J., Kühnert, D., Vaughan, T., Wu, C. H., Xie, D., ... & Drummond, A. J. (2014b). BEAST 2: A software platform for Bayesian evolutionary analysis, *PLoS Computational Biology*, 10(4), p. e1003537.

Buck, C. E., Cavanagh, W. G., Litton, C. D., & Scott, M. (1996). *Bayesian Approach to Interpreting Archaeological Data*, (Wiley), pp. 226–233.

Caussinus, H., Courgeau, D., & Mandelbaum, J. (2010). Estimating age without measuring it: A new method in paleodemography, *Population*, 65(1), pp. 117–144.

Didier, G., & Laurin, M. (2024). Testing extinction events and temporal shifts in diversification and fossilization rates through the skyline Fossilized Birth-Death (FBD) model: The example of some mid-Permian synapsid extinctions, *Cladistics*, 40(3), pp. 282–306.

Duchêne, S., Lanfear, R., & Ho, S. Y. (2014). The impact of calibration and clock-model choice on molecular estimates of divergence times, *Molecular Phylogenetics and Evolution*, 78, pp. 277–289.

Felsenstein, J. (2004). *Inferring Phylogenies* (Sinauer Associates).

Fernandes, R., Millard, A. R., Brabec, M., Nadeau, M. J., & Grootes, P. (2014). Food reconstruction using isotopic transferred signals (FRUITS): A Bayesian model for diet reconstruction, *PLoS One*, 9(2), p. e87436.

Gascuel, O., & Steel, M. (2020). A Darwinian uncertainty principle, *Systematic Biology*, 69(3), pp. 521–529.

Gavryushkina, A., & Zhang, C. (2020). Total-evidence dating and the fossilized birth-death model, *The Molecular Evolutionary Clock: Theory and Practice*, pp. 175–193.

Gigante, M., Mazzariol, A., Bonetto, J., Armaroli, E., Cipriani, A., & Lugli, F. (2023). Machine learning-based Sr isoscape of southern Sardinia: A tool for bio-geographic studies at the Phoenician-Punic site of Nora, *PLoS One*, 18(7), p. e0287787.

Giles, S., Feilich, K., Warnock, R. C., Pierce, S. E., & Friedman, M. (2023). A Late Devonian actinopterygian suggests high lineage survivorship across the end-Devonian mass extinction, *Nature Ecology & Evolution*, 7(1), pp. 10–19.

Greenhill, S. J., Drummond, A. J., & Gray, R. D. (2010). How accurate and robust are the phylogenetic estimates of Austronesian language relationships?, *PLoS One*, 5(3), p. e9573.

Greenhill, S. J., Haynie, H. J., Ross, R. M., Chira, A. M., List, J. M., Campbell, L., ... & Gray, R. D. (2023). A recent northern origin for the Uto-Aztecan family, *Language*, 99(1), pp. 81–107.

Greenhill, S. J., Heggarty, P., & Gray, R. D. (2020). Bayesian phylolinguistics, *The Handbook of Historical Linguistics*, 2, pp. 226–253.

Hastings, W. K. (1970). Monte Carlo sampling methods using Markov chains and their applications, *Biometrika*, pp. 97–109.

Heggarty, P., Anderson, C., Scarborough, M., King, B., Bouckaert, R., Jocz, L., ... & Gray, R. D. (2023). Language trees with sampled ancestors support an early origin of the Indo-European languages, *Science*, 381(6656), eabg0818.

Horvath, S. (2013). DNA methylation age of human tissues and cell types, *Genome Biology*, 14, pp. 1–20.

Kadkhodaie, Z., Guth, F., Simoncelli, E. P., & Mallat, S. (2023). Generalization in diffusion models arises from geometry-adaptive harmonic representation, arXiv preprint arXiv:2310.02557.

Lahtinen, M., Arppe, L., & Nowell, G. (2021). Source of strontium in archaeological mobility studies — Marine diet contribution to the isotopic composition, *Archaeological and Anthropological Sciences*, 13, pp. 1–10.

Lemey, P., Rambaut, A., Welch, J. J., & Suchard, M. A. (2010). Phylogeography takes a relaxed random walk in continuous space and time, *Molecular Biology and Evolution*, 27(8), pp. 1877–1885.

Lewis, P. O. (2001). A likelihood approach to estimating phylogeny from discrete morphological character data, *Systematic Biology*, 50(6), pp. 913–925.

Liu, X., Seguin-Orlando, A., Chauvey, L., Tressières, G., Schiavinato, S., Tonasso-Calvière, L., ... & Orlando, L. (2023). DNA methylation-based profiling of horse archaeological remains for age-at-death and castration, *(electronics): Iscience,* 26(3), 106144.

Louca, S., & Pennell, M. W. (2020). Extant timetrees are consistent with a myriad of diversification histories, *Nature,* 580(7804), pp. 502–505.

Łukasik, S., Bijak, J., Krenz-Niedbała, M., & Sinika, V. (2021). Paleodemographic analysis of age at death for a population of Black Sea Scythians: An exploration by using Bayesian methods, *American Journal of Physical Anthropology,* 174(4), pp. 595–613.

Luo, A., Zhang, C., Zhou, Q. S., Ho, S. Y., & Zhu, C. D. (2023). Impacts of taxon-sampling schemes on Bayesian tip dating under the fossilized birth-death process, *Systematic Biology,* 72(4), pp. 781–801.

Makarewicz, C. A., & Sealy, J. (2015). Dietary reconstruction, mobility, and the analysis of ancient skeletal tissues: Expanding the prospects of stable isotope research in archaeology, *Journal of Archaeological Science,* 56, pp. 146–158.

Matzke, M. N. J. (2013). Package 'BioGeoBEARS'. https://citeseerx.ist.psu.edu/document?repid=rep1&type=pdf&doi=70ee8c72dae97e1f65c25c269b706a dae378d41b.

Metropolis, N., Rosenbluth, A. W., Rosenbluth, M. N., Teller, A. H., & Teller, E. (1953). Equation of state calculations by fast computing machines, *Journal of Chemical Physics,* 21(6), pp. 1087–1092.

Michael, L., de Carvalho, F., Chacon, T., Rybka, K., Sabogal, A., Chousou-Polydouri, N., & Kaiping, G. (2022). Deriving calibrations for Arawakan using archaeological evidence, *Interface Focus,* 13(1), https://doi.org/10.1098/rsfs.2022.0049.

Moore, J. W., & Semmens, B. X. (2008). Incorporating uncertainty and prior information into stable isotope mixing models, *Ecology Letters,* 11(5), pp. 470–480.

Otárola-Castillo, E., Torquato, M. G., & Buck, C. E. (2023). The Bayesian inferential paradigm in archaeology, *Handbook of Archaeological Sciences,* 2, pp. 1193–1209.

Papamarkou, T., Skoularidou, M., Palla, K., Aitchison, L., Arbel, J., Dunson, D., ... & Zhang, R. (2024). Position: Bayesian deep learning is needed in the age of large-scale AI, *Forty-first International Conference on Machine Learning.* https://arxiv.org/abs/2402.00809.

Parnell, A. C., Phillips, D. L., Bearhop, S., Semmens, B. X., Ward, E. J., Moore, J. W., ... & Inger, R. (2013). Bayesian stable isotope mixing models, *Environmetrics,* 24(6), pp. 387–399.

Pedersen, J. S., Valen, E., Velazquez, A. M. V., Parker, B. J., Rasmussen, M., Lindgreen, S., ... & Orlando, L. (2014). Genome-wide nucleosome map and cytosine methylation levels of an ancient human genome, *Genome Research*, 24(3), pp. 454–466.

Ramsey, C. B. (2009). Bayesian analysis of radiocarbon dates, *Radiocarbon*, 51(1), pp. 337–360.

Resnik, P., & Hardisty, E. (2010). *Gibbs Sampling for the Uninitiated* (LAMP-153).

Ronquist, F., & Huelsenbeck, J. P. (2003). MrBayes 3: Bayesian phylogenetic inference under mixed models, *Bioinformatics*, 19(12), pp. 1572–1574.

Ronquist, F., Klopfstein, S., Vilhelmsen, L., Schulmeister, S., Murray, D. L., & Rasnitsyn, A. P. (2012). A total-evidence approach to dating with fossils, applied to the early radiation of the Hymenoptera. *Systematic Biology*, 61(6), pp. 973–999.

Tao, Y., Wei, Y., Ge, J., Pan, Y., Wang, W., Bi, Q., ... & Zhang, M. (2023). Phylogenetic evidence reveals early Kra-Dai divergence and dispersal in the late Holocene, *Nature Communications*, 14(1), p. 6924.

Truman, K., Vaughan, T. G., Gavryushkin, A., & Gavryushkina, A. (2024). The fossilised birth-death model is identifiable, bioRxiv, 2024-02.

Warnock, R. C., Yang, Z., & Donoghue, P. C. (2017). Testing the molecular clock using mechanistic models of fossil preservation and molecular evolution. *Proceedings of the Royal Society B: Biological Sciences*, 284(1857), https://doi.org/10.1098/rspb.2017.0227.

Wolpert, D. H., & Macready, W. G. (1995). *No Free Lunch Theorems For Search*, Technical Report SFI-TR-95-02-010, 10, 12. (Santa Fe Institute), pp. 2756–2760.

Wright, A. M., Bapst, D. W., Barido-Sottani, J., & Warnock, R. C. (2022). Integrating fossil observations into phylogenetics using the fossilized birth-death model. *Annual Review of Ecology, Evolution, and Systematics*, 53(1), pp. 251–273.

Wu, M. S., & List, J. M. (2023). Annotating cognates in phylogenetic studies of Southeast Asian languages. *Language Dynamics and Change*, 1, (aop), pp. 1–37.

Xu, Y., Li, X., Yang, Y., Li, C., & Shao, X. (2019). Human age prediction based on DNA methylation of non-blood tissues, *Computer Methods and Programs in Biomedicine*, 171, pp. 11–18.

Part 5

High in the Sky and Down to Earth:
Applications of Data Analytics

Chapter 7

Phylogenetics in Astrophysics

Didier Fraix-Burnet

University Grenoble Alpes,
CNRS, IPAG, Grenoble, France

7.1 Introduction

Phylogenetics is the study of the diversification of life on our planet Earth. Evolution takes the specific form of the Tree of Life explained by Darwin as driven by transmission with modification giving an advantage to the most adapted living organisms. Indeed, as we are looking for life elsewhere in the Universe, one of the most fundamental and general definition of life appears to be exactly this Darwinian process of evolution.

In our Universe, everything evolves, the Universe itself and all its constituents, whether it is stars, gas or galaxies. The essential difference with living organisms is that they do not replicate themselves (through sexuation or duplication) nor do they struggle for survival. Nevertheless, stars and galaxies somehow diversify through transmission with modification, opening the path toward a graphical representation of their evolution on trees or networks. This idea has been initially developed by Fraix-Burnet *et al.* (2006a, 2006b, 2006c) for galaxies and later applied to stars, stellar clusters and small bodies in our Solar System.

In this chapter, we first describe the most general of all phylogenetic methods, the Maximum Parsimony (also called cladistics) that is readily applicable to astrophysical entities. We then present what kind of evolution is found in astrophysics and give a few illustrative astrophysical results. All studies performed so far use individual objects considered as representatives of classes generally not properly defined. We thus introduce some current attempts to build the equivalent of "species" in astrophysics using the machine learning branch of classification. We end this chapter on the possible future for the depiction of the diversification of astrophysical entities.

7.2 Maximum Parsimony (Cladistics)

The hierarchical organization of biological diversity was found quite early in the Middle Ages. It led Linné to devise a nomenclature that is still used nowadays. This scheme of relationships between species was explained only in the 19th century by Darwin with his discovery of the evolution that creates diversity of species through a branching pattern. Another century was required for the corresponding methodology, called cladistics, to be devised by Hennig (1965). In this view, two (or more) species are related if they share a common history, that is if they possess properties inherited from a common ancestor. Cladistics is not concerned with genealogy (who is parent of whom) but with phylogeny (who is the cousin of whom). The objects under study are called taxa and represent species. Importantly, each node of the tree is not an observed taxon but an hypothetical ancestor because this ancestor is generally not available. Also, future discoveries may reveal other taxa that share this same ancestor or that are closer to it, thus modifying the branching structure around this node.

On the practical side, cladistics requires taxa to be described by characters. They are descriptors for which transformation can be documented, ideally with an ancestral state and a derived state. This means that these characters have kept a trace of the historical evolution of the different species. They can however be affected by redundancies, incompatibilities, too much variability (reversals), parallel and convergent evolutions, that tend to destroy the tree structure.

Algorithmically, cladistics has been associated in the 80's to the search of a maximum parsimony tree that minimizes the total number of state changes. This tree depicts the simplest evolution scenario given the data set (Felsenstein, 1984). The total number of state changes corresponds, after labeling of the internal nodes, to the sum over all edges of the absolute difference between the values of the characters between two successive nodes. The maximum parsimony approach can be applied to continuous characters or values through discretization.

The drawback is that the analysis must consider all possible trees before selecting the most parsimonious one. Since this is rapidly impossible when the number of taxa or character states increases, the tree with the minimum score is searched for with some heuristics (Felsenstein, 1984). The advantage is that the maximum parsimony algorithm, like all character-based techniques, can take uncertainties or unknowns into account. This is very useful in astrophysics, where measurements uncertainties are unavoidable and missing data common. This is of course not possible with distance-based methods.

When several equally most parsimonious trees are found, a consensus tree is built. Each node of this tree is valued with the percentage of occurrence of this node among all the most parsimonious trees. This provides an estimation of the robustness of the various branching patterns, but the reliability of the tree is best estimated by a bootstrap analysis.

7.3 The Phylogeny of Stars and Galaxies

7.3.1 *The evolution of the universe*

Similarly as to living organisms on Earth, the diversification of astrophysical objects is intimately linked to the evolution of their environment. After the phase of the Big Bang, some 13.7 Gyr ago, the Universe was relatively simple: very homogeneous with tiny density fluctuations, that later grow by gravitational attraction and lead to the formation the first stars and galaxies. The Universe was initially composed with 75% of hydrogen and 25% of helium, with traces of some other light elements such as lithium. All the other chemical elements (oxygen, carbon, iron, nitrogen...) are yielded by nuclear fusion within stars. During the first billions years, the collisions between galaxies and clusters of galaxies are very frequent, creating diversity in shapes, dynamics, and chemical mixtures. Most of the stars formed during this phase. Later, interactions between galaxies occurred only within clusters and more rarely.

This schematic history of astronomical objects explains why phylogenetic tools have been successfully applied on stars and galaxies. At a smaller scale and for different reasons, small bodies of our Solar System also constitute good targets for maximum parsimony.

7.3.2 *The life cycle of stars*

Stars are formed from the collapse of a molecular cloud found in the inter-stellar medium. A molecular cloud is made of gas and dust grains. Its chemical composition is specific and depends on its history. When the most massive stars explode, they release the heavy chemical elements such as carbon, oxygen or iron that they produced, henceforth modifying the chemical composition of the interstellar medium from which new stars will be formed. In our Galaxy for instance, three populations are tradition-ally identified (populations I, II, and III) according to the fraction of heavy atoms.

The evolution of stars is thus characterized by a cycle through which the gas transformed in the core of massive stars is transmitted to new generations that inherit a modified chemical composition. Despite that the new stars are not direct off-springs of the previous generation, this process is very similar to the diversification of species with a Darwinian mecha-nism of transmission with modification, pointing to the possibility to build tree-like graphs to depict the evolution of stellar populations.

Similarity in the chemical composition of stars is not enough to gather stars that are born in the same molecular cloud since they evolve depend-ing on their mass and on their age. As a consequence, stars of the same family do not form clumps in the space of chemical composition, but rather diverging evolutionary paths. Blanco-Cuaresma and Fraix-Burnet (2018) showed that phylogenetic methods such as Maximum Parsimony can solve the chemical tagging goal, that aims at identifying families of stars that were born at the same place at the same time, and were subse-quently dispersed throughout our Galaxy. Using the distance-based Neighbor-Joining Tree approach, Jofré *et al.* (2017) and Walsen *et al.* (2024) tackle large samples of stars to understand the evolution of our Milky Way and the origin of the stellar populations.

7.3.3 *Globular Clusters*

Globular clusters are large ensemble of stars that are tightly gravitation-ally bound, more or less looking as a small spherical galaxy without much interstellar gas. As such, it is possible to study the populations of their stars. In the globular cluster Ω Centauri, the phylogenetic tree (Fraix-Burnet and Davoust, 2015) reveals seven groups that correspond to differ-ent origins for the molecular clouds that formed the stars.

The globular clusters in our Galaxy can also be studied as individual taxa (Fraix-Burnet *et al.*, 2009). The tree reveals three groups that correspond to three populations of globular clusters that formed during three stages in the assembly of our Galaxy.

7.3.4 *Evolution of galaxies*

Galaxies are complex ensembles of billions of stars, gas, dust grains, and also black holes of different masses that are all gravitationally bound. If we consider the single property of morphology, two spiral galaxies that merge create a new object, most generally a galaxy of elliptical shape. This new galaxy obviously contains the material from its progenitors (transmission) but with modified properties such as the kinematics or new generations of stars (modification). There are many other ways a galaxy can change some of its properties. There can be external perturbations (such as a close encounter with another galaxy) or internal turbulence (supernova explosions, nuclear activity...). In all case one can identify a transmission with modification mechanism. Obviously, the physics of the evolution of galaxies is more complex than that of stars because of its many constituents and interactions. This complexity is somehow not that different from the complexity of living organisms.

Indeed, there is a well-known hierarchical tree-like representation of our evolving Universe that is called the merger tree of dark matter haloes (see e.g., Stewart *et al.* 2008). The driver of their diversification is the gravitational attraction that leads the haloes to merge and grow in mass and size. The merger tree is a kind of genealogic tree that relates the very first overdensities of dark matter to the largest concentration found nowadays. The root of the merger tree is the largest haloes, that is the most diversified and recent ones. From a phylogenetic point of view, a cladogram would appear in a reverse orientation, with the very first tiny overdensities as the root and the species of bigger haloes occurring progressively by successive mergers of smaller ones. Since galaxies are supposedly formed within these dark matter haloes, the formation and evolution of galaxies is often thought as a linear hierarchical scenario, while observations suggest a more complex diversification scheme.

A lot of studies have demonstrated that galaxies can be understood on a phylogenetic diagram. The first application of maximum parsimony in astrophysics used galaxies computed in cosmological simulations

(Fraix-Burnet *et al.*, 2006b, 2006c). The advantage is that the true evolution of all the galaxies is known. A complementary work on real dwarf galaxies (Fraix-Burnet *et al.*, 2006a) completed the demonstration of the physical relevance of phylogenetic representation of galaxy diversification.

Many studies followed and mapped the diversification of relatively small samples of galaxies (Fraix-Burnet *et al.*, 2010, 2012, 2017, 2019; Martínez-Marín *et al.*, 2020; de Lima *et al.*, 2023). A synthesis of the phylogeny of clades based on about one thousand galaxies was presented in (Fraix-Burnet *et al.*, 2012) with the identification of the physical processes that trigger the branching patterns in the diversification of galaxies.

To really grasp the complexity of the physics of the diversification of galaxies, considerably larger samples of galaxies should be considered. For this, we should first establish a proper multivariate classification to define species. This is the subject of the next section.

7.3.5 *Small objects in our solar system*

Holt *et al.* (2018, 2021) applied cladistics on small objects in our Solar System. The idea of family comes from the fact that collisions between large objects create swarms of smaller ones that keep essentially the same chemical and physical properties of their parent, and share some similar dynamics. In the case of the satellites of Jupiter and Saturn (Holt *et al.*, 2018) identified new families of retrograde and irregular objects. For the Jovian Trojan swarms (small objects that share Jupiter's orbit; Holt *et al.*, 2021), 48 clans are identified, indicating groups of objects that possibly share a common origin. Several clans contain members of the known collisional families but the clans are often broken into subclans, and most can be grouped into 10 superclans, reflecting the hierarchical nature of the population. An interesting use of such work is the selection of particularly significant objects for future space missions.

7.4 "Species" in Astrophysics

All phylogenetic studies in astrophysics so far used individual objects as representative of classes. This approach is suited to small families of objects, such as small bodies of our solar system or chemical tagging of stars, but cannot encompass the complexity of the huge population of galaxies across the cosmic times. We need to identify the classes, or the equivalent of species in biology, to group together similar objects.

7.4.1 *Classification in astrophysics*

Astronomy (understood as the observation of the sky) is a very old activity, but astrophysics (understood as the description of the objects) is a relatively young discipline as compared to biology. The reason is simply because stars are too far away, not to speak of galaxies, so Humanity had to wait for some technological inventions to identify the physical properties of astronomical entities. As a consequence, the traditional visual classification technique remained sufficient until the end of the 20th century and the advent of the Big Data era in astronomy. The main component of the traditional classification is the use of a particular trait to build and name a classification. For instance, color and intensity (called magnitudes) are obvious criteria. It happens that they are closely related to temperature of stars for the first one and mass of galaxies for the second one, thus reinforcing their apparent usefulness.

Classification in stellar astrophysics is slightly more developed because stars are simpler and appear much brighter than galaxies in our sky, hence more easily caught by our instruments. There are many classes defined by some observational characteristics (red giant, brown dwarf, low mass, variable, cataclysmic, main sequence, neutron...), evolutionary stage (classes 0, 1, 2, pre-main sequence, novae . . .) or chemical (population I, II, III, ...). The spectra of stars are extremely rich, but the seven main classes (O, B, A, . . .) are essentially based on their color. Sub-classes were added depending of the properties of some emission or absorption lines of some chemical elements. Each class is characterized by an observed typical spectrum. They cannot however be qualified of multivariate classifications from a statistical point of view.

For galaxies, there is no such taxonomic classification apart from general types based on conspicuous observational or physical traits. The most famous example is the morphological classification first established by Hubble and later refined by de Vaucouleurs. The spectra of galaxies are also classified by eye using some particular features. Because of the complexity of galaxies, numerous and sometimes incompatible categories exist. For instance, elliptical galaxies are supposedly red and not forming stars, but there are many exceptions.

Astronomers are using sophisticated approaches to classify automatically observations in data bases. This kind of supervised approach requires very large training samples of known and labeled classes. For images of galaxies, these samples are built through the citizen science project Galaxy Zoo, but they remain limited in size and depend on visual inspection with the associated human biases. This is probably why Deep Learning techniques are somewhat disappointing. They can reach high levels of accuracy (80–90%) but either on a small number of classes or on detailed features present in the images (Khramtsov *et al.*, 2022). Indeed, supervised machine learning algorithms appear to focus on the detection of detailed features that are often not identified in the de Vaucouleurs 18-class scheme (Fraix-Burnet, 2023).

In astrophysics we are thus very far from the achievements in biology. This may have been sufficient up to now, but the diversity of observed astronomical entities is becoming so huge that we are faced with the same questions as the biologists in the 17th century. Distance analysis was invented at that time to create a multivariate objective classification of living organisms. Nowadays, this technique would be called unsupervised classification (or clustering) with machine learning. Unsupervised learning is the art of asking the algorithm to look for structures in the data space, it is up to the scientist to give a physical meaning to these structures. Biologists use distance-based similarity analyses to define species, this is called phenetics.

Unsupervised techniques appeared less than 15 years ago in astrophysics, and gained a lot more interest when unsupervised Deep Learning (autoencoders) were invented around 2015. Some fascination for foundation models in artificial intelligence is leading the exploration of new pathways, with some hidden dream that the machine will do all the work for us! Unfortunately, the unsupervised approach is inherently difficult because the interpretation of the results are not given. But is this not the main interest of research?

In this section, we present the most developed path toward an unsupervised classification of galaxies. It uses a statistical technique (sometimes called "classical machine learning") that is relatively easy to understand.

7.4.2 *Unsupervised Machine Learning*

7.4.2.1 *Model-based clustering*

Clustering divides a given data set $Y = \{y_1, ..., y_n\}$ of n data points, described by p variables, into K homogeneous groups. A popular clustering technique uses Gaussian mixture models, which assumes that each class is represented by a Gaussian probability density. The data are therefore modeled by a density $f(y, \theta) = \Sigma^K_{k=1} \pi_k \varphi(y, \theta_k)$ where φ is a p-variate normal density with parameter $\theta_k = \{\mu_k, \Sigma_k\}$ containing the means μ_k and the covariance matrices Σ_k of the Gaussian distribution for class k. The parameters π_k are the mixing proportions. The K-means method described elsewhere in this book is a particular case that assumes that the covariance matrices $\Sigma_k = \Sigma$ are diagonal (thus corresponding to spherical distributions) and identical for all the clusters.

7.4.2.2 *The algorithm Fisher-EM*

The algorithm Fisher-expectation-maximization (Fisher-EM) (Bouveyron and Brunet, 2012) is a Gaussian mixture model algorithm in a discriminant latent subspace. It uses a modified version of the EM optimizing loop by inserting a Fisher-step that optimizes the ratio of the sum of the between-class variance over the sum of the within-class variance, hence optimizing the Gaussian mixture and the subspace together for a better clustering. In Fisher-EM, the clustering process occurs in the subspace of dimension $d < p$, thus performing a dimensionality reduction adapted for the clustering. Therefore, the Gaussian mixture model is applied to the projected data X rather than the observed data Y: $Y = UX + \varepsilon$, where U is the projection matrix and ε is a noise vector of dimension p following a Gaussian distribution centered around 0 and of covariance matrix Ψ. The observed data is now modeled by the marginal distribution $f(y) = \Sigma^K_{k=1} \pi_k \varphi(y; U\mu_k, U\Sigma_k U^t + \Psi)$. The parameters of the model (π_k, U, Σ_k, Ψ, and the number K of clusters) are obtained through the optimization of a likelihood, thus providing an objective criterion to decide for the best solution.

7.4.3 *Application to spectra of galaxies*

The algorithm Fisher-EM has been applied on several samples of galaxies with several kinds of sets of variables. We here describe only its use on spectra of galaxies, because the number of variables is very high (up to five thousand) and the algorithm is especially designed for the high-dimension. In addition, the spectrum of galaxy contains nearly all the information we can get from the light that reaches us.

The first robust unsupervised classification of spectra of galaxies has been performed on 702,248 spectra of galaxies (Fraix-Burnet *et al.* 2021). 86 classes are found, each with a typical spectrum and little dispersion within the class. The 37 most populated ones gather 99% of the sample, the other classes being either rarer objects or spectra affected by instrumental defects. The size of the sample gives a good idea of the diversity of galaxies in the near Universe, and could yield the equivalent of "species" needed for phylogenetic studies of galaxies.

An extension to this work considered about 80,000 spectra of galaxies up to redshift of 1.4 (Dubois *et al.*, 2024). Because of this redshifting effect, it is not possible to compare all the spectra at once since there is no common variable between the closest and the farthest galaxies. The only way is to perform a clustering at each epoch and compare the classes between two subsequent epochs. A tree can thus be constructed, but this is a chronological tree, not a phylogenetic one. Pavlou *et al.* (2023) made a similar temporal tree from clustering on graphs made by similarity of spectra.

7.5 Perspectives

One may assume that we are on a good track to define species of galaxies from a multivariate similarity point of view. Then, what do we need to build a phylogeny? Characters. We need variables that can provide the evolutionary stages of the diversification of galaxies. Biologists have long used morphometric descriptors to define and describe species. The evolution of living organism is also driven by the genes. Biologists struggled a long time to find that the ribosomal RNA gene sequences is the right entity to build a phylogeny. It appears that the phylogenies obtained from the two kinds of characters are in very good agreement.

For stars, the chemical composition seems a good candidate and has proven to be rather effective. Somehow, it can be considered as the equivalent of the DNA. For galaxies, chemistry is obviously important, but we most often have access only to the properties of the stars and the interstellar medium averaged over the galaxy. Mass is an interesting indicator since gravity dictates a nearly systematic growth of galaxy masses. Colors are an indicator of the average age of the stellar populations, but this hides the real star formation history, and color is also affected by dust. These properties might correspond to morphometric descriptors, but they are only a few. Much more information is contained in spectra, but it is still not clear what could constitute the equivalent of DNA. Spectra could be a source of characters, but it is still difficult to identify the right variables.

Interactions and mergers add another difficulty in the galaxy diversification scenario. They create hybridization that are known to alter the tree-like structure of the phylogeny. In biology, it is still possible to represent the diversification in the form of reticulograms or even split networks. Networks are a more general representation of evolutionary relationships than trees. They are also, more difficult to read and to use for a physical interpretation.

Maximum Parsimony will probably remain a reference because of its conceptual and practical relative simplicity to implement. Neighbor Joining Tree technique is also straightforward to use. Probabilistic methods via maximum likelihood optimization are very effective in biology but require some model for the evolution of characters. The basic physics of the astronomical objects and their evolution is in principle well known. However, we have not yet identified the evolutionary characters, nor do we have proposed any probabilistic model to describe the complexity of the detailed physical processes at play in galaxies.

It is difficult to evoke the future without mentioning artificial intelligence, even if this term is broad and vague. The advent of unsupervised learning triggered many new kinds of data exploration, clustering being one important subject. Astronomers realize however that the machine cannot provide everything, and the question of the interpretability becomes central. A lot of investigations are going on in foundation models, with the more or less hidden hope that if we put everything we know about a galaxy, then an artificial intelligence algorithm could be able to answer some question a scientist may have. I am personally very skeptical after my own experience with the interpretation of groupings found by a fully understood algorithm. The science of classification, systematics, teaches us that a classification should comprise a name and a full description. Can we ask an artificial intelligence algorithm to invent names for us and provide a description of something we have never seen? We have many powerful algorithms to find clusters in a data space but we still need a lot of work to define a classification. Finding evolutionary paths using a phylogenetic approach is very recent in astrophysics and it is only the beginning of an adventure that may, in a remote future, extent the tree of life to the origin of the Universe!

References

Blanco-Cuaresma, S., & Fraix-Burnet, D. (2018). A phylogenetic approach to chemical tagging. Reassembling open cluster stars, *Astronomy and Astrophysics*, 618, p. A65.

Cardone, F., & Fraix-Burnet, D. (2013). Hints for families of GRBs improving the Hubble diagram, *Monthly Notices of the Royal Astronomical Society*, 434(3), pp. 1930–1938.

de Lima, M., Porpino, K., & da Silva, J. (2023). Astrocladistics: Evolutionary classification for galaxies of the M81 group, *Astrophysics and Space Science*, 368(4), p. 29.

Dubois, J., Siudek, M., Fraix-Burnet, D., & Moultaka, J. (2024). From VIPERS to SDSS: Unveiling galaxy spectra evolution over 9 Gyr through unsupervised machine-learning, *Astronomy and Astrophysics*, 687, p. A76.

Felsenstein, J. (1984). The statistical approach to inferring evolutionary trees and what it tells us about parsimony and compatibility, *Cladistics: Perspectives on the Reconstruction of Evolutionary History* (Columbia University Press, New York), pp. 169–191.

Fraix-Burnet, D. (2023). Machine learning and galaxy morphology: For what purpose? *Monthly Notices of the Royal Astronomical Society*, 523(3), pp. 3974–3990.

Fraix-Burnet, D., Chattopadhyay, T., Chattopadhyay, A., Davoust, E., & Thuillard, M. (2012). A six-parameter space to describe galaxy diversification, *Astronomy and Astrophysics*, 545, p. A80.

Fraix-Burnet, D., Choler, P., & Douzery, E. (2006a). Towards a phylogenetic analysis of galaxy evolution: A case study with the dwarf galaxies of the local group, *Astronomy and Astrophysics*, 455, pp. 845–851.

Fraix-Burnet, D., Choler, P., Douzery, E., & Verhamme, A. (2006b). Astrocladistics: A phylogenetic analysis of galaxy evolution I. Character evolutions and galaxy histories, *Journal of Classification*, 23, pp. 31–56.

Fraix-Burnet, D., & Davoust, E. (2015). Stellar populations in Ω Centauri: A multivariate analysis, *Monthly Notices of the Royal Astronomical Society*, 450(4), pp. 3431–3441.

Fraix-Burnet, D., Davoust, E., & Charbonnel, C. (2009). The environment of formation as a second parameter for globular cluster classification, *Monthly Notices of the Royal Astronomical Society*, 398, pp. 1706–1714.

Fraix-Burnet, D., Douzery, E., Choler, P., & Verhamme, A. (2006c). Astrocladistics: A phylogenetic analysis of galaxy evolution II. Formation and diversification of galaxies, *Journal of Classification*, 23, pp. 57–78.

Fraix-Burnet, D., Dugué, M., Chattopadhyay, T., Chattopadhyay, A., & Davoust, E. (2010). Structures in the fundamental plane of early-type galaxies, *Monthly Notices of the Royal Astronomical Society*, 407, pp. 2207–2222.

Fraix-Burnet, D., Marziani, P., D'Onofrio, M., & Dultzin, D. (2017). The phylogeny of quasars and the ontogeny of their central black holes, *Frontiers in Astronomy and Space Sciences*, 4.

Fraix-Burnet, D., Mauro D'Onofrio, M., & Marziani, P. (2019). Maximum parsimony analysis of the effect of the environment on the evolution of galaxies, *Astronomy & Astrophysics*, 630, p. A63.

Holt, T., Brown, A., Nesvorny, D., Horner, J., & Carter, B. (2018). Cladistical Analysis of the Jovian and Saturnian Satellite Systems, *The Astrophysical Journal*, 859, p. 97.

Holt, T., Horner, J., Nesvorny, D., King, R., Popescu, M., Carter, B., & Tylor, C. (2021). Astrocladistics of the Jovian Trojan Swarms, *Monthly Notices of the Royal Astronomical Society*, 504(2), pp. 1571–1608.

Jofré, P., Das, P., Bertranpetit, J., & Foley, R. (2017). Cosmic phylogeny: Reconstructing the chemical history of the solar neighbourhood with an evolutionary tree, *Monthly Notices of the Royal Astronomical Society*, 467(1), pp. 1140–1153.

Khramtsov, V., Vavilova, I. B., Dobrycheva, D. V., Vasylenko, M. Yu., Melnyk, O. V., Elyiv, A. A., Akhmetov, V. S., & Dmytrenko, A. M. (2022). Machine learning technique for morphological classification of galaxies from the SDSS. III. Image-based inference of detailed features, *Space Science & Technology*, 28(5), pp. 27–55.

Martínez-Marín, M., Demarco, R., Cabrera-Vives, G., Cerulo, P., Leigh, N., & Herrera-Camus, R. (2020). A phylogenetic analysis of galaxies in the Coma Cluster and the field: A new approach to galaxy evolution, *Monthly Notices of the Royal Astronomical Society*, 499(4), pp. 5607–5622.

Pavlou, O., Papadopoulou Lesta I. M. V., Papadopoulos, M., Papaefthymiou, E. S., & Efstathiou, A. (2023). Graph theoretical analysis of local ultraluminous infrared galaxies and quasars, *Astronomy and Computing*, 45, p. 100742.

Walsen, K., Jofré, P., Buder, S., Yaxley, K., Das, P., Yates, R., Hua, X., Signor, T., Eldridge, C., Rojas-Arriagada, A., Tissera, P., Johnston, E., Aguilera-Gómez, C., Zoccali, M., Gilmore, G., & Foley, R. (2024). Assembling a high-precision abundance catalogue of solar twins in GALAH for phylogenetic studies, *Monthly Notices of the Royal Astronomical Society*, 529(3), pp. 2946–2966.

Chapter 8

Data Analytics for Tracing Methods of Ancient Astronomy

Susanne M Hoffmann[*,†] **and Boshun Yang**[*]

*University of Science and Technology of China,
Dept. of the History of Science and Scientific Archaeology,
Hefei, China*

†*Friedrich Schiller University of Jena,
Faculty of Mathematics and Computer Science,
Jena, Germany*

8.1 Intro: The Dataset, the Ancient Star Catalogs

On the examples of historical star catalogs from Europe and China, we elaborate on various methods of data analysis for the history studies.

8.1.1 *Ptolemy's star catalog*

From Greco-Babylonian antiquity, only one star catalog is preserved: the star catalog of Claudius Ptolemy in a book called Μαθηματικὴ Σύνταξις (*Scientific Compendium*),[1] 137 CE. This book preserves a lot of practical astronomical knowledge, e.g., how to build scientific instruments to measure the stars, store the data and compute predictions of

---

[1] Edition: (Heiberg, 1898), current standard translation: (Toomer, 1984).

(heliacal) rising and setting times. In his description of how to make a celestial globe,[2] Ptolemy provides a dataset to plot the stars on a solid sphere (Alm. VII,1 to VIII,3). A globe maker needs three facts per star: the coordinates (longitude, latitude) and the magnitude of the template to choose. These three numbers are given in the Almagest. Of course, the magnitudes are a measure of the brightness of the star, but to our knowledge, there is no scientific device to measure these brightnesses. These estimates of magnitudes may depend on the vicinity or on the star's color, as human cognition is easily affected by the surroundings of a specific reception (Protte and Hoffmann, 2021). Grasshoff (2021) suspects that Ptolemy might have measured the magnitudes of stars in twilight observations. Still, no method description is preserved from antiquity. Hoffmann (2022) pointed out that mere estimates might be processed for the visual appearance of the globe.

Greek and Babylonian predecessors of Ptolemy's star catalog do not preserve magnitudes as a measure of brightness, but they only have textual descriptions. It is essential to distinguish the magnitudes of the stars on a visual map to enable people to recognize patterns (constellations) for orientation. The orientation serves not only the instrument's user but also the globe maker and the author of the book: As Ptolemy groups the stars in his catalog according to constellations, he also describes their position in the figure. The schema of the text is, therefore:

Headline (constellation name)

Position of the star in the constellation figure , longitude, latitude, magnitude.

Ptolemy's star catalog is the only one preserved in this schema. Hipparchus (2nd century BCE) is said to have made a similar, unpreserved catalog. It is scholarly speculated that Hipparchus's star catalog was no longer copied because it was merged into that of Ptolemy. As Ptolemy's coordinates predate his book by roughly a century (Fig. 8.1), it is assumed that he computed his data from earlier observations with a wrong precession constant. Ptolemy was even falsely accused of stealing Hipparchus's data (Newton, 1977). He undoubtedly took over some

---

[2]Globes are used for computations at that time, so they are important devices.

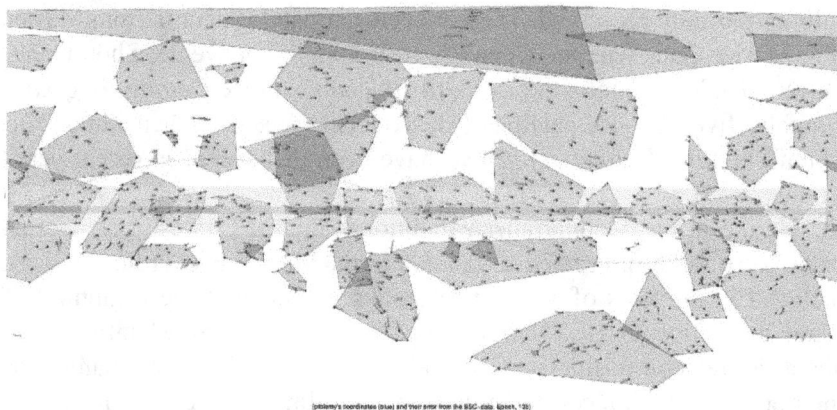

Fig. 8.1. The polygons indicate Ptolemy's constellations in ecliptic coordinates. The points are the star positions given in his catalog, and their "tails" indicate their real position (computed for 137 CE). All positions are systematically shifted to the west. The median[3] of this error is minimized for 48 CE (Hoffmann, 2017: 39).

data (Grasshoff, 1990). To address this question of the origin of Ptolemy's data, Hipparchus's catalog is reconstructed from the only document of his hand, a commentary[4] to the poem of Aratus.[5] In cases when Aratus mentions a star rising or setting or positioned with wrong data, Hipparchus criticizes the mistake for his student and gives the correct data (to his view) in an appendix. Hence, Hipparchus' dataset is necessarily incomplete (a) because he follows Aratus and comments only on mistakes and (b) because his appendix exclusively refers to rising and setting constellations and not to the ever-visible ones.

8.1.2 *Chinese star catalogs*

It is difficult to ascertain the total number of star catalogs produced in ancient China. As of now, we can confirm that there have been more than 13 traditional ancient Chinese star catalogs (excluding those influenced by European or Islamic astronomy) in the epochs from the 2nd century BCE to the 14th century CE. Among these, 11 catalogs have either been

---

[3] The "median" is an average that excludes outliers, and therefore throws out writing errors.
[4] Standard translation still is the German one by Manitius (1894).
[5] Phaenomena, English standard translation by Kidd (1997).

fully or partially preserved. These eleven star catalogs come from Master Shi, Master Gan, Wuxian, Chen Zhuo, Yixing, Yang Weide, Zhou Cong, Ouyang Fa, Yao Shunfu, the *Tianwen Huichao*, and the *Datong Tongzhan*. Notably, five of these catalogs (Master Gan, Wuxian, Chen Zhuo, Yao Shunfu and the *Datong Tongzhan*) have been discovered in recent years (Yang, 2023).

Unlike in the Mediterranean traditions, ancient Chinese star coordinates were rarely presented in tabular form. Most star coordinates were included in the texts of astrological books, typically at the beginning of astral divination texts. These catalogs primarily contained information about the relative positions of constellations or asterisms, star names (or their positions within constellations) and coordinates. The coordinates are the lunar lodge degrees (*ruxiudu*, 入宿度), which are the difference in right ascension (RA) between the observed star and the determinative star of its lunar mansion, and the polar distance degrees[6] (*qujidu*, 去极度). If the constellation were the Twenty-Eight Lunar Mansions, the first coordinate would be the mansion width degrees (*xiudu*, 宿度). These numbers represent the difference in RA between adjacent determinative stars. Some Chinese star catalogs also included the declination. Before the Tang Dynasty (618–907), some star catalogs also incorporated the distances of stars to the ecliptic along the direction of RA, which is called by Yabuuti (1937) as "Polar Celestial Latitude" or "Pseudo Celestial Latitude".

Chinese observers typically measured only the coordinates of the brightest or the rightmost stars within constellations, so there is one coordinate per constellation. Due to the high number of constellations in ancient China, the discovered star catalogs contain data for about three hundred stars at most. According to Sun and Kistemaker (1997: 53), many determinative stars in the Master Shi's star catalog were in the western part of their constellation and carried characters like "先至 *Xianzhi*" ("coming first"). This indicates the method of observation, namely while transiting the meridian. In the Song Dynasty's Huangyou star catalog, the stars of the Celestial Market Enclosure (Tianshi Yuan) were arranged in

---

[6] In mathematics, the separation from the pole in spherical coordinates is generally referred to as the "polar angle" θ because it is counted from the pole. In astronomy, the equatorial latitude is called "declination" and counted from the celestial equator to the poles. In historical frames of reference, when a declination-like coordinate is counted from the pole, it is sometimes called "co-declination" and sometimes (equivalently) "polar distance (degrees)".

RA order from west to east rather than being listed in their north-south order according to its shape, which stresses the conclusion on meridian observations.

Unfortunately, traditional Chinese star catalogs didn't quantify stellar magnitudes. However, there is evidence that they had some descriptive language for star brightness, such as "大星" (large stars), "明者" (bright ones), and "小星" (small stars) (Chen, 2008). Most extant star maps do not distinguish between the brightness of stars, either. In the rare instance of the *Cheonsang Yeolcha Bunyajido* map, differences in brightness were depicted. Yet, it does not accurately reflect their brightness (e.g., Sirius is depicted smaller than Arcturus). These distinctions were likely based on the impression of the mapmaker rather than precise systematic observations.

It is noteworthy that before the Tang Dynasty, the number of star catalogs was quite limited and that equatorial coordinates were often used for centuries with minimal changes after their initial measurements. However, in the Tang Dynasty, astronomers realised that previous coordinates didn't fit the location of stars of their time, and the production of star catalogs began to accelerate rapidly. This phenomenon was particularly evident during the Northern Song period, with new star catalogs being created every few decades or even within no more than 20 years.

8.2 Data Analysis

8.2.1 *Hipparchus: Reconstruction of lost data*

Hipparchus's scholarly commentary on Aratus' didactical poem comes in three parts, among which the first one discusses Aratus's verse individually, the second gives Hipparchus's own, very schematic view of risings and setting, and the third part is a list of hour stars. Hour stars are arranged at lines of the same RA and, this way, provide this coordinate directly. The computations of declinations (and more RAs) from the data in the second part are a bit more sophisticated.

In general, the ecliptic longitude λ of any star differs from its simultaneously culminating point λ_c at the ecliptic. However, all simultaneously culminating stars share the same RA with λ_c. That means Hipparchus's system of defining a star's position is equatorial (it uses the meridian as the hand of the celestial clockwork). Still, he uses the scale at the circle of the ecliptic to enumerate the RAs or hour angles he provides. As Hipparchus works with declinations,[7] but in his description of measured star positions, he speaks about an "ecliptic longitude in an equator parallel" (Manitius,1894: 151; Hoffmann, 2017: 623). Hoffmann (2017: 630) concludes that he had a non-orthogonal coordinate system in mind. Like all globes at this time, Hipparchus's celestial globe probably carried the equator, the ecliptic and the tropics — either drawn or as built mechanical bands surrounding the globe. Additionally, it had equator parallels drawn and ecliptic longitudes marked on its surface.

The schema of Hipparchus's text is very strict:

> Star X in Constellation Y rises/ sets while (simultaneously)
> Ecliptic degree λ_H is rising/ setting, and
> Ecliptic degree λ_C is culminating and
> Star1, Star2, Star3 are culminating.

Any culmination, whether a star or a point of the ecliptic, is a position at the meridian and has, thus, a well-defined hour angle, which in turn translates to a well-defined RA according to the fundamental equations of spherical trigonometry that can be found in any school book of astronomy:

(1) All simultaneously culminating points (stars or ecliptic degrees) share the same RA. Therefore, the RA is known for all stars mentioned in culmination because the RA of the ecliptic degree is computable with a simple coordinate transform.
(2) The currently culminating RA equals the local sidereal time θ. Hence, when a star is given in rising and setting, the duration of its time above the horizon is provided by the difference of the two given local sidereal times $\theta_{set} - \theta_{rise}$, a measure for the azimuth A of the rising/ setting.
(3) The duration above the horizon measures the star's declination DE. It only depends on the geographical latitude, which was normalized $\varphi = 36°$ North for all measurements in Greek antiquity.

---

[7]No matter whether he measures them from the pole, from one of the tropics or from the equator: In any case, the underlying concept is the "latitude in the equatorial system" that today is called declination.

This way, we can compute the coordinates of the stars from Hipparchus's data (azimuth A, inclination ϵ, geographical latitude φ):

$$RA_{\text{culm}} = \arctan \frac{\cos \epsilon \sin \lambda - \sin \epsilon \tan \beta}{\cos \lambda} \text{ (with } \beta = 0 \text{),} \qquad (1.1)$$

$$RA_{\text{hor}} = \theta - \arctan \left(\frac{\tan A}{\sin \varphi} \right), \qquad (1.2)$$

$$DE = \arcsin(-\cos \varphi \cos A). \qquad (1.3)$$

The declination is computable if and only if the star is given at different positions so that the sidereal time interval is computable (Vogt, 1925; Grasshoff, 1990; Hoffmann, 2017). Therefore, Hipparchus's coordinates are only completely given for a small fraction of the stars he mentions in his text (62–67, depending on the identification of star names).

Once the coordinates of Hipparchus's source (in this case, his globe with which he wrote the educational text that we call "commentary") are computed, we can compare them to Ptolemy's coordinates and modern measurements of these positions.

8.2.2 *Did Ptolemy copy Hipparchus's data?*

8.2.2.1 *Accuracy and format of Hipparchus's data*

The second and third parts of Hipparchus' commentary deal with time measurements. No declinations or ecliptic latitudes are listed. These coordinates are included only in the first part, where Hipparchus directly criticises Aratus's wording, which makes it clear that he uses an equatorial frame of reference (not ecliptic coordinates like Ptolemy). The latitude-like coordinate does not have a clear reference: among the 43 numbers of this type, 21 (roughly half) are given as pole distance, 14 (38%) are given as distance from the equator (now called declination), and eight (18%) are given as distance to the tropical circles. This shows that he had the concept of parallel circles to the celestial equator but no convention to measure them from only one basic line (equator) or point (pole) as we do today. He did not have the concepts of declination or co-declination.

Furthermore, Hoffmann (2017: 201, "Befund 2.47") analysed all datasets in Hipparchus's commentary separately and found that the declinations have an average error of 0.1° while the RAs have an average error of 1° (Hoffmann, 2017: 121, "Befund 2.15" and 203, "Befund 2.49"). The fact that the average error of declinations is much smaller (10%) than the average error for the longitude-like coordinate (Hoffmann, 2017) suggests that the two coordinates were measured with different methods: the declinations were measured with an angular instrument, while the RAs were measured with a clock (which was easier).

8.2.2.2 *Accuracy of Ptolemy's data*

Vogt (1925) and Grasshoff (1990: 84–91) discuss the fractions of degrees given in Ptolemy's Almagest to determine the accuracy of Ptolemy's instrument. One could estimate that the smallest fraction listed in the catalog is either the smallest unit of the scale or the middle between two points at the scale and, thus, half of the smallest unit. As Ptolemy's numbers of arcminutes are always divisible by five, Vogt (1925) and Grasshoff (1990) derive the smallest unit of the instrument of 5′ or 10′ (five or ten arcminutes).

Concerning the overall distribution of the fractions of degrees (Fig. 8.2), one could get the impression that this instrument had a more accurate scale in latitudes than in longitudes. The data provide values with a regular grid of ten arcminutes (0′, 10′, 20′, 40′ with 20% each, and 30′ and 50′ with 10% each) in longitudes, whereas latitudes are given with a clear preference for 0′ and 30′ (full and half degrees: 43% of the data points).

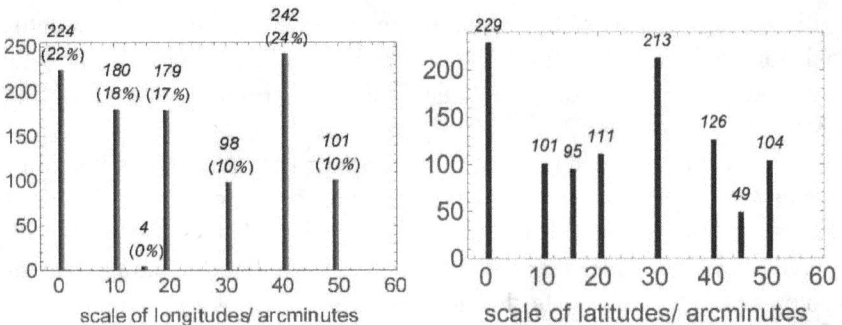

Fig. 8.2.   Fractions of degrees on the scales of Ptolemy's instrument.

However, a closer look at the distribution of the values of numbers reveals different patterns (Fig. 8.2). In longitude, the middle between 10 and 20 arcminutes (15′) is given in only four cases (0.4% of 1028 values). Also, among the 1028 coordinates, 49 latitudes are given with a fraction of 45 arcminutes (¾ degree) and 95 latitudes are given with a fraction of 15 arcminutes (¼ degree). The numbers of 5′, 25′, 35′ and 55′ never occur but only 15′ and 45′ (one quarter and three-quarters of a degree) in latitude, while in longitude, only the quarters of degrees.

These patterns in the distribution of fractions suggest that Ptolemy's instrument (or the instrument of his predecessors who he quoted) had two different scales or that even two different instruments were used for measuring longitude and latitude. One of his datasets might have been measured with an instrument with a scale of 0.5° as the smallest unit in latitude and 15 and 45 arcminutes as the smallest fraction (half of the smallest unit) to be read. Another one of his sources might have contributed measurements with an instrument whose smallest unit at the latitude scale was 10′ (ten arcminutes).

The rather equal distribution of fractions in longitude suggests that they were measured with the same (type of) instrument. This may be a hint to interpret Ptolemy's statement that he measured the positions of the stars as righteous:

> *Hence, again using the same instrument (...), we observed as many stars as we could sight down to the sixth magnitude.* (Alm VII, 4, cited from Toomer 1984, 339)

He may have partially re-measured coordinates catalogued in the previous centuries. Still, as RAs can be measured with clocks instead of angular instruments, Ptolemy remeasured all ecliptic longitudes with his new armillary sphere instead of computing longitudes from earlier measured RAs. His instrument might have been incorrectly calibrated which caused the systematic offset (Vogt, 1925).

Due to Ptolemy's remeasurement of all longitudes but only some of the latitudes, his data is more accurate in longitude than in latitude. In contrast, Hipparchus's data was more accurate in latitude than longitude, for he had measured RAs with a clock and declinations with an angular instrument (Hoffmann, 2017: 204, 207). Ptolemy may have taken many latitudes directly from Hipparchus, for his error bar is estimated to 1° (Hoffmann, 2017: 57, 121). Grasshoff's key witness of common errors in

Ptolemy's and Hipparchus's datasets, π Hydrae, with a latitude[8] error of 4.86° (Grasshoff, 1990: 97), confirms the hypothesis that mainly latitudes were copied from Hipparchus to Ptolemy.[9]

Anyway, the smallest unit of the instrument's scale is not automatically the error bar (or uncertainty of the measurement) because the uncertainty of the measurement also depends on statistical parallax scattering while reading and on possible error propagation in cases of computations and calibrations. The procedure for using the armillary sphere for measurements as described in Alm. V, 1 requires frequent calibration during the night.[10] Subsequently and as derived from star catalog data, the error bars of the values given in the Almagest are much bigger than the smallest unit (Hoffmann, 2017: 56).

From the modern point of view, we can compare Ptolemy's data for the star positions with satellite data (of course, correcting the well-known deviation of Ptolemy's longitudes to the equinox in his century) and determine the average error of his measurements. The resulting estimated error bars of the latitudes and longitudes in the Almagest are roughly one degree (cf. scatter ellipses in Grasshoff (1990: 189, 192–194) and chapter in Hoffmann (2017: 56) "Befund 2.1").

8.2.2.3 *Ptolemy's source(s)*

As Ptolemy's accuracy exceeds Hipparchus's one in the commentary, these two texts do not share the same source (or only very indirectly). Hipparchus's text is written on the desk with a globe whose smallest unit was 1°. This globe preserved the data from measurements (with two

---

[8] The big error of this star in latitude and the fact that nobody corrected it in centuries might be explainable by the function of the Hydra-star chain for globe makers: it marked the celestial equator (cf. analysis of the Kugel Globe in Hoffmann, in print) but π Hydrae is a bit more south than the other stars in the chain.

[9] Grasshoff's other witness of transfer, θ Eri, with a big error in longitude might be explained with a zero point problem, either in writing or in the adjustment of the instrument because this is the only star of this constellation west (and not east) of 0° Ari. Ptolemy's other big errors, ι CMa, 81 Leo, φ UMa, and α Mic can be explained with writing errors (Hoffmann, 2017: 45).

[10] Practical exercise described in Evans (1987a, 1987b) and Evans (1998: 255), Ptolemy's armillary rebuild by Wlodarczyk (1987) and Protte and Hoffmann (2021: 137–141) with test measurements.

different instruments) but possibly not accurately as the globe might have been too small. Ptolemy's star catalog was written for globe makers: it preserves processed data with the intention to be plotted on globes (Alm. VIII, 3). His instrument (the armillary sphere) had two scales (longitude and latitude) with equal units of 10 arcminutes, so dense that it was impossible to read half units. Yet, the raw data stems from at least two different sources, as shown above. Hoffmann (2018) presents the same conclusion based on the fact that there are stars in Hipparchus's dataset that are not in the Almagest star catalog. Hoffmann (2017: 106)[11] had already pointed out that Hipparchus's biggest errors are not repeated in the Almagest, so the data has been reworked in between. Nevertheless, as Grasshoff (1990) found common errors in the Almagest and in Hippachus, Hipparchus's data might have been one of the sources of Ptolemy, but not the only source (Hoffmann, 2018).

8.2.3 China: Calculation of the observing epoch of stellar coordinates

Chinese historians face other challenges in determining who observed these data and when. Occasionally, astral divination texts containing star catalog data include the observer's name and the observation date. However, such catalogs that provide the original observational information are rare, and they might even be forged and need careful analysis. Furthermore, the observation dates for most star catalog data lack explicit textual records, necessitating age calculations and other clues to pinpoint the observation period. Over the past century, historians of astronomy have continuously tried to employ new mathematical methods to calculate the observation dates of ancient star catalogs, applying various statistical methods to refine the dating and reduce error margins.

The most characteristic case in this field is Master Shi's star catalog. This catalog contains over 100 star coordinates, including the lunar lodge degrees, the polar distance degrees, and the position and distance of stars relative to the ecliptic. These coordinates are derived from a book called Master Shi's Star Canon, attributed to Shi Shen or Shi Shenfu, a renowned astronomer believed to have lived during the 4th century of the Warring States period (476 BCE–221 BCE). However, Shi Shen's exact birth and

---

[11] Hoffmann (2017: 106) "Befund 2.12", arguments on the previous pages.

death dates are unclear, and there are doubts about whether he personally observed the data in this catalog. Consequently, many scholars have performed calculations on this star catalog. Since the North Pole slowly shifts due to the precession of the Earth's axis, the polar distances for each star also change over time. Therefore, the observation date can be determined based on the recorded polar distances in the star catalog.

The Master Shi's star catalog has undergone countless transcriptions throughout history, inevitably leading to some errors in the data. One of the earliest researchers of this catalog, Ueta (1930), considered this aspect. His method aimed to avoid interference from erroneous data in dating calculations. He drew circles centred on the measured stars with the recorded polar distances as radii. In this map, the arcs of correct data from the same period would converge almost at the same point (Fig. 8.3), representing the North Pole of the observation date. This method helps minimize the impact of transcription errors or significant observational inaccuracies on the results. His findings indicated two clusters of intersection points: one around the North Pole of 360 BCE and another around 200 CE. He concluded that the catalog contains data from observations made in two different periods; therefore, the star catalog was initially observed by Shi Shen himself, but due to the likely loss of some parts over time, subsequent generations supplemented it with additional observations. This hypothesis explained the presence of data from different periods within the same star catalog, reflecting a mix of two epochs. Recently, He and Zhao (2025) obtained similar results using a similar method, the generalized Hough transform.

Unfortunately, while this method considers the impact of outlier data on the results, it overlooks the systematic errors inherent in the instruments themselves. If the instrument has a certain degree of systematic error, each recorded co-declination value will carry this error. Given that the northern celestial pole height of the used armillary sphere was 36 du (365.25 du = 360°) compared to the true north pole at 34.8 du, this star catalog exhibits a systematic error of about 1°. This systematic error causes all observed co-declination values to be approximately 1° smaller than the theoretical values. Due to the precession of the Earth's axis, the celestial north pole moves in a specific direction. The co-declination values of stars on the hemisphere (A) in the direction of motion gradually decrease, while those on the opposite hemisphere (B) increase. Due to the observed co-declination values being smaller than the theoretical values, the calculation of observation dates for stars in hemisphere A appears later

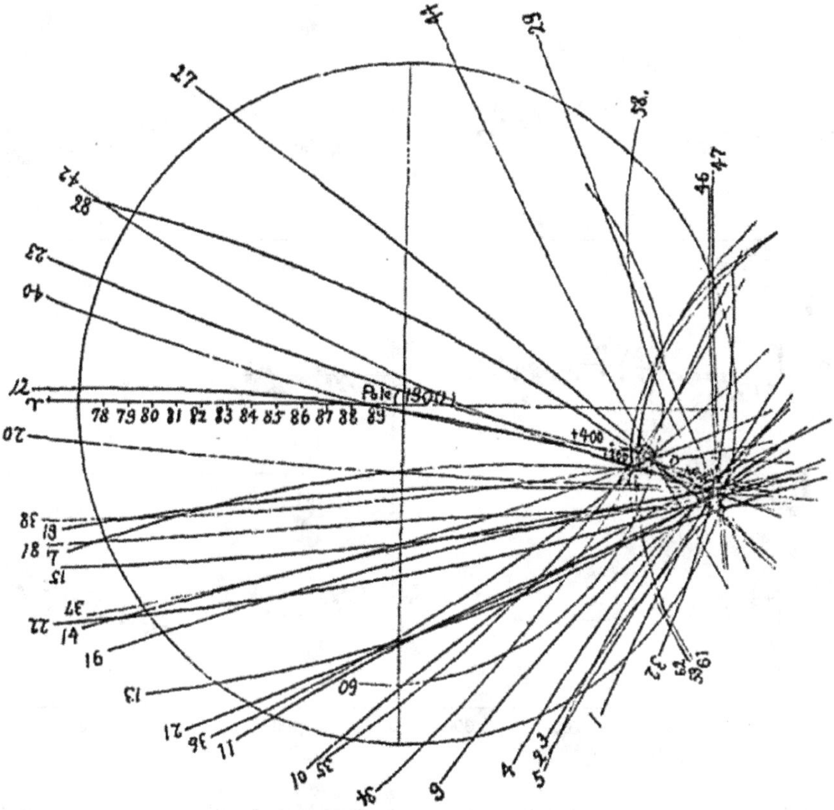

Fig. 8.3. Ueta's graphical method showed two clusters of intersection points (from Ueta (1930, Fig. 5)).

than the true observation dates. For stars in hemisphere B, the calculated dates appear earlier than the true dates.

Maeyama (1975, 1976) found the systematic error in this star catalog and accounted for this factor in his calculations. He set a series of possible observation epochs with fixed time intervals and calculated the standard deviation of the observational errors for each date. By finding the date with the minimum standard deviation of the errors, he determined the observation date to be around 60 BCE. This approach of setting multiple dates and using the least squares method to determine the true date was adopted by later scholars. A notable example is Sun and Kistemaker (1997: 37–74), who were the first to use Fourier analysis (Fig. 8.4) to

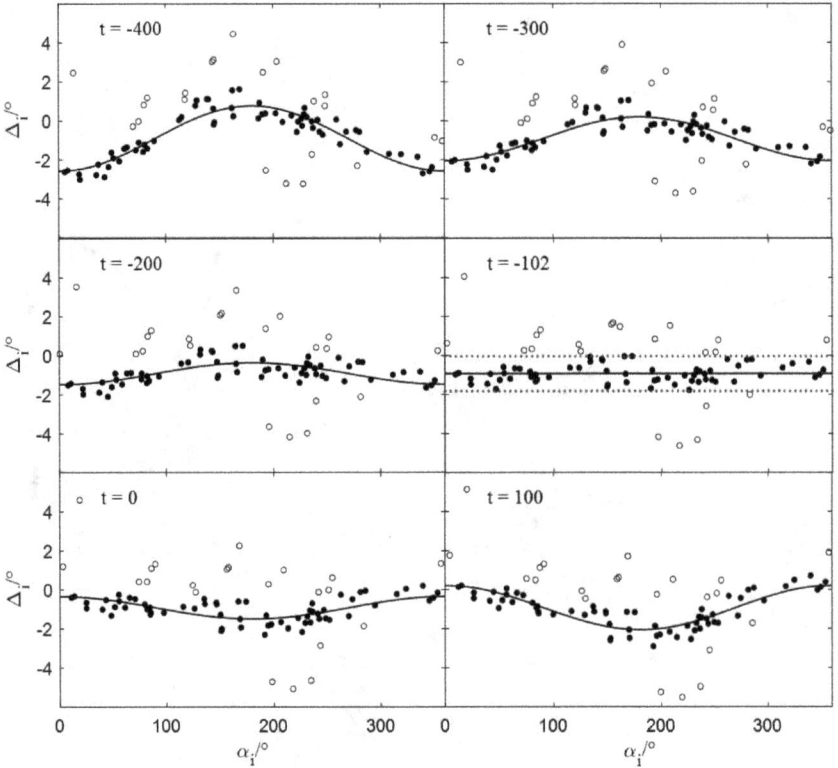

Fig. 8.4. Fourier fitting curve of declination errors of Master Shi's star catalog (without the 28 lunar mansions). In the fourth picture, the solid points within the dashed lines represent the data used for calculations, with a standard deviation of s; the hollow points represent the data eliminated after iteration, with absolute values greater than 2s, and are not used for dating calculations.

determine the observation date of Master Shi's star catalog. This method effectively illustrates the distribution of systematic bias and observational errors, providing a more visual representation of the data.

The Fourier analysis method involves fitting a cosine function $f(x) = c + a \cos \alpha$ (where c represents the systematic bias and $a = n \times (t_s - t_r)$, with $n = 0.5567°$/century is the rate of declination change) to the differences Δ_i between the theoretical declination values δ_i^{cal} and the observed values δ_i^{obs} for a given assumed date t_s. The closer the assumed date t_s is to the actual observation date t_r, the smaller the coefficient a of the cosine term in the fit function, resulting in a flatter curve. When the coefficient a

is zero, the assumed date equals the observation date, and the curve is horizontal. Using this method, Sun and Kistemaker (1997: 64) determined the observation date to be 78 BCE.

However, statistical methods like Fourier analysis combined with least squares are not without flaws, as they are susceptible to the influence of extreme or erroneous values. When determining the date of a star catalog, most scholars assume that the dataset they are processing originates from a single period. In reality, data from a star catalog may contain observations from other epochs or even fabricated data. An iterative filtering method can be employed to address this issue.

First, the observation date t_r and systematic bias c are obtained after performing a Fourier fit on the errors Δ_i. The random errors are isolated after removing the systematic bias from the errors. Then, the data with random errors greater than twice the standard deviation s_Δ are excluded. This process of calculating the standard deviation and excluding large errors is repeated with the remaining errors iteratively for j times until the remaining random errors are all within the range of twice their standard deviation s_Δ^j. Then, the date t_r' indicated by the remaining data is calculated, and all random errors for this date are filtered again to obtain a new standard deviation s_Δ^j. This cycle continues until the result for the data within twice the s_Δ^j range remains stable, which is taken as the resulting standard deviation s and observation date t_r.

Yang (2023: 40–42) refined this method based on the research of Anh (2020). Using this approach, the observation date for Master Shi's star catalog was determined to be around 103 BCE ± 15, which coincides with the time of the calendar reform and astronomical observation activities during the Taichu period (104 BCE–102 BCE) of the Western Han Dynasty (202 BCE–8 CE).

8.2.4 *What we know from Master Shi's star catalog*

The Chinese sky culture comprises 283 constellations with 1,464 stars, organized by Chen Zhuo in the third century CE. Master Shi's star catalog records over a hundred of the brighter constellations, forming the primary framework of the Chinese celestial system. This catalog is one of the Chinese constellation system's most significant sources and components. The constellations recorded in it represent China's oldest and most important constellations, but the time of their formation is debated. Shi Shen, who is credited in the catalog, lived during the Warring States period of

China. Calculations by scholars like Ueta seemed to correspond to Shi Shen's era, leading them to claim that the catalog is the world's oldest star catalog. However, later research revealed that Ueta's calculations overlooked some systematic errors, suggesting that the catalog is a product of the Western Han period, not the direct observations of Shi Shen.

Current calculations link Master Shi's star catalog to the calendar reforms during the Taichu era. Historical records indicate that during the calendar reform, the emperor assembled a large group of astronomers from the civilian sector to conduct observations and make algorithmic improvements. This group included Tang Du and Luoxia Hong. Tang Du (fl. 2nd century BCE) was an astrologer whose constellation system and astrological theories had been adopted by the royal astronomers Sima Tan and his son Sima Qian before the reform. Luoxia Hong, on the other hand, was the inventor of the Chinese armillary sphere and an important algorithmic innovator. Emperor Wu appointed Tang Du for the research area called "Section of Heaven", while Luoxia Hong was responsible for the calendar calculations. The exact business of "Section of Heaven" is unclear, but it likely pertained to astrological work related to constellations. Another clue shows that Tang Du created a star map based on the coordinates of stars measured by the armillary sphere. Therefore, it is highly likely that Tang Du collaborated with Luoxia Hong during this astronomical reform to observe and record the positions of the stars. Although the star catalog is attributed to Shi Shen, the observers might have been Tang Du and Luoxia Hong, along with other astronomers from the Western Han period.

By comparing Master Shi's star catalog with Tang Du's older constellations, we can discern the specific work Tang Du carried out during the reform. The constellations recorded in Master Shi's star catalog clearly connect to Tang Du's older constellation system, which Sima Tan recorded. Still, they also highlight distinct differences, indicating that Tang Du made certain adjustments to the constellations during this astronomical reform. The most significant adjustment likely pertains to the northern polar constellation. In Sima Tan's records, the "North Pole" constellation comprised only four stars, while Master Shi's star catalog lists five stars in the same constellation. This change appears to be driven by practical observational needs. To accurately measure the positions of the stars, particularly their polar distances, it was essential first to align the armillary sphere accurately with the north pole star adopted at that time. During the early Western Han period, astronomers continued using the polar star from the Shang Dynasty (1600 BCE–1046 BCE), β UMi, even though by that time it was already 8° away from the celestial north pole. The 1° error in Master Shi's star catalog suggests that Tang Du and Luoxia Hong conducted new measurements to the north pole and discovered that the old pole star was too far from the true north pole and consequently added a star closer to the northern pole to the "North Pole Constellation", making it the fifth star in this constellation.

This catalog provides details about the instruments used during that period. The data are typically read to a precision of 1/4 du. The rare phrase "a little more/less than 1/4'" indicates that the armillary sphere used at that time might have been divided into four segments per du. However, this finer scale did not effectively improve observational accuracy because the errors from other aspects of the armillary sphere were much greater than the precision of the finer divisions. According to calculations, 14.1% of the data has random errors exceeding 2°, and 26% of the data has errors exceeding 1°. This suggests that the armillary spheres of the Western Han period were not yet highly refined. The eccentricity of the rings might be relatively high, and the scales might not have been evenly spaced. In contrast, star catalogs from the mid-Song Dynasty onwards show almost no random errors exceeding 2°, indicating significant improvements in the precision of astronomical instruments by that time. This makes it clear that the manufacturing technology of metal armillary spheres, which only recently appeared during the Western Han Dynasty, was still not very mature.

8.2.5 *Conclusion*

Data analysis helps a lot by reconstructing the past. We can easily compute missing data and reconstruct historical instruments and sources by analyzing the data given in Hipparchus's and Ptolemy's texts. By comparing the reconstructed or preserved ancient data with modern measurements (e.g., in the case of stellar positions), we can also determine the errors of ancient data. By tracking the errors, we can also determine the ways of knowledge transfer. The identification and analysis of error types can provide insights into the structural details of the instruments at that time, offering reliable corroboration or supplements to historical records.

Foremost, we conclude that reconstructing the date of Mediterranean star catalogs uses the position of the equinoxes. In contrast, the date of

East Asian star catalogs is determined by the position of the celestial (equatorial) north pole. Copying processes and methods of observations and computations can be reconstructed by error analysis in various sources. In both cultures, we determine the questions of who observed the raw data and when by analysing the given numbers and their error bars.

As star catalogs are datasets ready for data analysis, this has been done for centuries, even before the invention of electronic computers. However, the new technical IT solutions provide more possibilities for derived knowledge from comparing errors in different catalogs and comparing more datasets simultaneously.

References

Ahn, S. (2020). Revisiting the epoch of the earliest Chinese star catalog titled "Shi Shi Xing Jing". *PASJ*, 72(5), pp. 1–21.

Chen, J.-J. (2008). *Zhongguo Gudai Tianwenxuejia* 中国古代天文学家 (*Ancient Chinese Astronomers*). (Zhongguo kexue jishu chubanshe, Beijing), pp. 35–37.

Evans, J. (1987a). On the Origin of the Ptolemaic Star Catalogue. *Journal for the History of Astronomy 18*, Part One (August 1987a), pp. 155–172.

Evans, J. (1987b). On the Origin of the Ptolemaic Star Catalogue. *Journal for the History of Astronomy 18*, Part Two (November 1987b), pp. 233–278.

Evans, J. (1998). *The History and Practice of Ancient Astronomy* (Oxford University Press, New York/Oxford).

Grasshoff, G. (1990). *The History of Ptolemy's Star Catalogue* (Springer, New York).

Grasshoff, G. (2021). Measurements of Stellar Magnitudes in Ptolemy's Almagest, *Proceedings of the Splinter Meeting in the Astronomische Gesellschaft*, ACHA, pp. 147–151.

He, B.-L., & Zhao, Y.-H. (2025). Determining the observational epoch of the Shi's star catalog using the generalized Hough transform method. Accepted by *Research in Astronomy and Astrophysics*. arXiv:2504.02186.

Heiberg, J. L. (1898). *Claudii Ptolemaei Syntaxis Mathematica* (Teubner Verlagsgesellschaft, Leipzig).

Hoffmann, S. M. (2017). *Hipparchs Himmelsglobus — Ein Bindeglied in der babylonisch-griechischen Astrometrie?* (Reihe: Research, Springer Verlag, Wiesbaden/ New York).

Hoffmann, S. M. (2018). The Genesis of Hipparchus' celestial globe, *Mediterranean Archaeology and Archaeometry*, 18(4), pp. 281–287.

Hoffmann, S. (2022). Essay: On Ptolemy's Stellar Magnitudes, Hoffmann, S. M. and Wolfschmidt, G., (eds.), *Astronomy in Culture — Cultures of Astronomy* (tredition/ OpenScienceTechnology, Hamburg/ Berlin), pp. 426–429.

Hoffmann, S. M. (2024). Some results on the ancient globes, *Proceedings of the Symposium of the International Coronelli Society in Berlin, Globe Studies*.

Hunger and Steele (2019). *The Babylonian Astronomical Compendium MUL. APIN* (Routledge, Abingdon, Oxfordshire, UK).

Kidd, D. (1997). *Aratus: Phaenomena*, Cambridge Classical Texts and Commentaries, Series Number 34, Illustrated Ed. (Cambridge University Press).

Maeyama, Y. (1975). On the astronomical data of ancient China (ca-100 ——+200), A numerical analysis. *Archives Internationales d'Histoire des Sciences*, 25, pp. 247–276.

Maeyama, Y. (1976). On the astronomical data of ancient China (ca-100 ——+200), A numerical analysis. *Archives Internationales d'Histoire des Sciences*, 25, pp. 27–58-JM.

Manitius, C. (1894). *Hipparchos: Kommentar zu den "Himmelserscheinungen" des Aratos und des Eudoxos* (Teubner Verlag, Leipzig).

Newton, R. R. (1977). *The Crime of Claudius Ptolemy* (Johns Hopkins University Press, Baltimore and London, USA).

Protte, P., & Hoffmann S. M. (2020). Accuracy of magnitudes in pre-telescopic star catalogues, *Astronomische Nachrichten*, 341, pp. 827–840.

Protte, P., & Hoffmann. S. M. (2021). Pre-telescopic star catalogues — Accuracy in magnitudes and positions, Wolfschmidt, G. and Hoffmann, S. M. (eds.), *Applied and Computational Historical Astronomy* (tredition/OpenScienceTechnology, Hamburg/Berlin), pp. 137–141.

Sun, X.-C., & Kistemaker, J. (1997). *The Chinese Sky during the Han* (Leiden: Brill), p. 53.

Toomer, G. J. (1984). *Ptolemy' Almagest* (Princeton University Press, 1998, Orig.: London).

Ueta, J. (1930). Shih Shen's catalogue of stars, the oldest star catalogue in the orient. *Publications of the Kwasan Observatory* (Kyoto Imperial University), 2(1), pp. 17–48.

Vogt, H. (1925). Versuch einer Wiederherstellung von Hipparchs Fixsternverzeichnis, *Astronomische Nachrichten*, 224(2), pp. 17–56.

Wlodarczyk, J. (1987). Observing with the Armillary Astrolabe. *Journal for the History of Astronomy*, 18(3), pp. 173–195.

Yabuuti, K.薮内清 (1937). 唐开元占经中の星经 (The Star Catalogue in the "Tang- K'ai-yüan-chan-ching" (唐开元占经)). *The Toho Gakuho*京都：东方学报: *Journal of Oriental Studies, Kyoto*, 8, pp. 56–74.

Yang, B.-S.杨伯顺 (2023). *Zhongguo Chuantong Hengxing Guance Jingdu ji Xingguan Yanbian Yanjiu* 中国传统恒星观测精度及星官演变研究 (A Research on the Accuracy of Chinese Traditional Star Observation and the Evolution of Constellations), PhD thesis, (Hefei: University of Science and Technology of China, 2023) pp. 40–41.

Chapter 9

Why Folktales Have "Node" Limits

Julien d' Huy

Laboratoire d'Anthropologie sociale, Collège de France,
CNRS, EHESS, Paris, France

9.1 Introduction

In 1903, Marcel Mauss (1903: 245–246) was struck by the limited variety of episodes and their combinations in mythic constructions: "There are only a few types of myths, and within the multitude of examples of each type, the parts appear in a roughly consistent order." (*il y a des types de mythes, en petit nombre, et dans la multitude des exemplaires de chaque type, les parties se présentent dans un ordre à peu près constant.*)

In oral literature, the term "tale type" refers to a narrative model composed of episodes and motifs arranged in a relatively stable manner. This is primarily a tool for classifying and studying oral tradition narratives, particularly through the Aarne–Thompson–Uther (ATU) classification system. This system has proven extremely valuable in structuring the wide diversity of folktales, thereby facilitating their analysis. It has also enabled the creation of extensive national and international catalogs (Le Quellec and Sergent 2017: 1306–1307).

The evolution of some of these tales has been modeled using tools borrowed from evolutionary biology by comparing the number of mythems — i.e., minimal semantic units — that different versions have in common. Myth trees have thus been constructed, and the most probable proto-versions reconstructed. Similarly, trees have been generated based on the distribution of motifs across different cultural areas. This latter approach takes into account the capacity of mythological motifs to recombine with one another (d'Huy, 2012a, 2023a, 2023b; d'Huy *et al.*, 2023). However, neither approach fully explains the variability of the data, and the use of networks, which highlight borrowings between versions and independent inventions (d'Huy, 2012b), does not perfectly capture certain evolutions.

Returning to folktales, a challenge posed to the phylomythological approach is that their forms are rarely "pure": it is common for tales not to belong strictly to a single type but to combine several. Vladimir Propp observed that "fairy tales have a peculiar feature: the parts constituting one story can be transferred without any change to another story" ("Сказки обладают одной особенностью: составные части одной сказки без всякого изменения могут быть перенесены в другую," Propp, 1928: 14–15). Similarly, Lutz Röhrich stated that "contamination is the essence of all folk poetry" ("Kontamination ist das Wesen aller Volksdichtung," 1976: 291). This contamination also appears among early editors of folktales, who did not hesitate to combine two or three versions of the same tale to produce a more complete version (Belmont, 2013). However, this practice is sometimes perceived, perhaps unconsciously, as an alteration of the original form (Dundes, 1969). François-Marie Luzel highlights this point when commenting on a version of "The Mermaid and the Hawk": "I reproduce it faithfully as I collected it, to give an idea of how certain storytellers, believing they are increasing the interest of their narratives, alter and mix them at will. The longer a tale is and the more full of marvels and trials, the more successful it usually is with winter evening audiences" (Luzel, 1887, II: 418; *Je le reproduis fidèlement tel que je l'ai recueilli, pour donner une idée de la manière dont certains conteurs, croyant augmenter l'intérêt de leurs récits, les altèrent et les mélangent, à plaisir. Plus un conte est long et rempli de merveilles et d'épreuves, plus il a de succès, ordinairement, auprès de l'auditoire des veillées d'hiver*).

Contamination in the transmission of folktales has been the subject of numerous studies. On this point, several articles in the Enzyklopädie des Märchens are worth consulting, such as "Affinität" (Voigt, 1977), "Assoziation" (Fischer and Lüthö, 1977), and "Kontamination" (Kawan, 1996).

Could a statistical analysis of these contaminations among different tale types be envisaged, allowing for a transition from macro-level analysis (phylomythology) to micro-level analysis, or simply to another level of analysis? Would it be possible, based on a corpus, to identify cycles or specific kinship links? And, if so, are there rules governing these recombinations?

9.2 The Fox's Web

To address these questions, I began in 2018 to study the relationships between folktales categorized under "The Clever Fox (other animals)" in the first volume of Types of International Folktales (Uther, 2011). This group primarily consists of short narratives in which a stronger protagonist is confronted by a cleverer adversary. For this analysis, I adopted a network approach, aiming to create a map of connections where tale types, represented as network nodes, are linked by edges if they are connected.

Before this study, network analysis had been used in diverse fields, such as social networks (e.g., Mac Carron and Kenna, 2012), biological systems (Proulx *et al.*, 2005), and the evolution of networks specific to certain tale types (Karsdorp and van den Bosch, 2016).

It is important to recall that a tale type is an abstract category encompassing similar narratives rather than a single story. This classification is imperfect: some versions may be misclassified, and the definition of a tale type remains broad. Some types describe specific narrative actions (e.g., ATU 1: "The Theft of Fish"), while others cover general narrative structures that include multiple tale types (e.g., ATU 6: "Animal Captor Persuaded to Talk"). Moreover, non-European traditions are significantly underrepresented in this database, a limitation acknowledged by Uther himself. Thus, Uther's work remains primarily a tool for textual and bibliographic research, and our results are only applicable to Western Eurasia.

The concept of a combination between tale types is complex to define. According to Uther, related narratives are those that belong to narrative cycles or that combine and "contaminate" each other. A mention of "combination" in his work indicates at least three occurrences of similar combinations among the tales studied. This minimizes classification errors but does not always distinguish between local cultural specificities and broader associations.

To avoid sampling biases, I analyzed high (at least eight examples) and low (fewer than eight examples) levels of combination between tale types, treating all nodes equally. Indeed, the distinction between seven and eight combinations depends on too many variable factors to be meaningful.

In our corpus of 33 selected tale types, each was connected on average to 3.79 others, suggesting the presence of recurring narrative cycles. An adjacency matrix was constructed and analyzed using Social Network Visualizer 2.3 (Kalamaras, 2017).

I calculated two centrality measures to evaluate the importance of each ATU: degree centrality (DC) (number of direct connections; Fig. 9.1) and closeness centrality (average geodesic distance between a node and all others; Fig. 9.2). ATUs 1, 2, 3, 4, 5, and 15 emerged as the most central. The average distance between two ATU tales was 2.97.

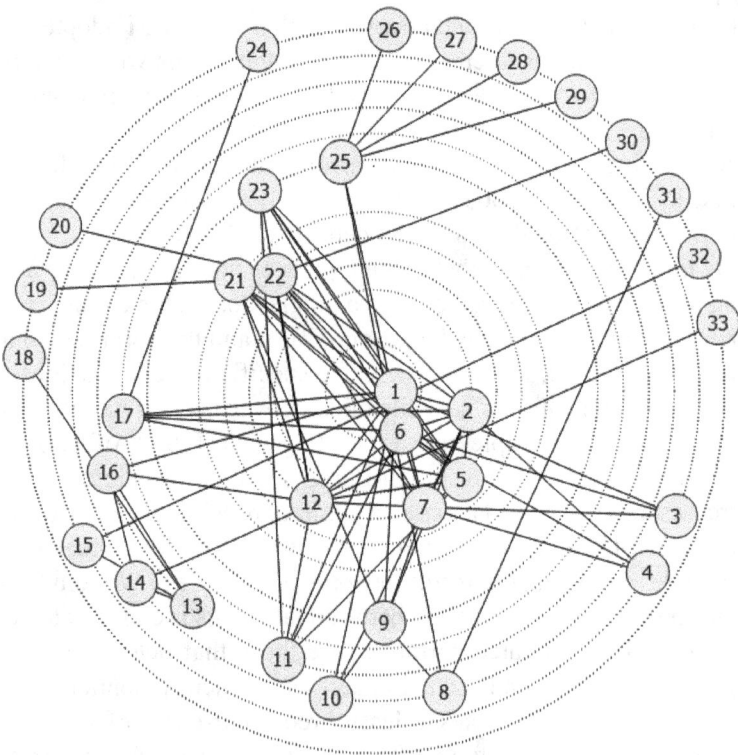

Fig. 9.1.   DC calculated from the corpus. Correspondence: 1/ ATU 1; 2/ ATU 2; 3/ ATU 2B; 4/ ATU 2D; 5/ ATU 3; 6/ ATU 4; 7/ ATU 5; 8/ ATU 6; 9/ ATU 8; 10/ ATU 8*; 11/ ATU 9; 12/ ATU 15; 13/ ATU 20A; 14/ ATU 20C; 15/ ATU 20D*; 16/ ATU 21; 17/ ATU 30; 18/ ATU 31; 19/ ATU 32; 20/ ATU 33; 21/ ATU 34; 22/ ATU 41; 23/ ATU 47A; 24/ ATU 47D; 25/ ATU 56A; 26/ ATU 56B; 27/ ATU 56D; 28/ ATU 56A*; 29/ ATU 57; 30/ ATU 60; 31/ ATU 61; 32/ ATU 62; 33/ ATU 65.

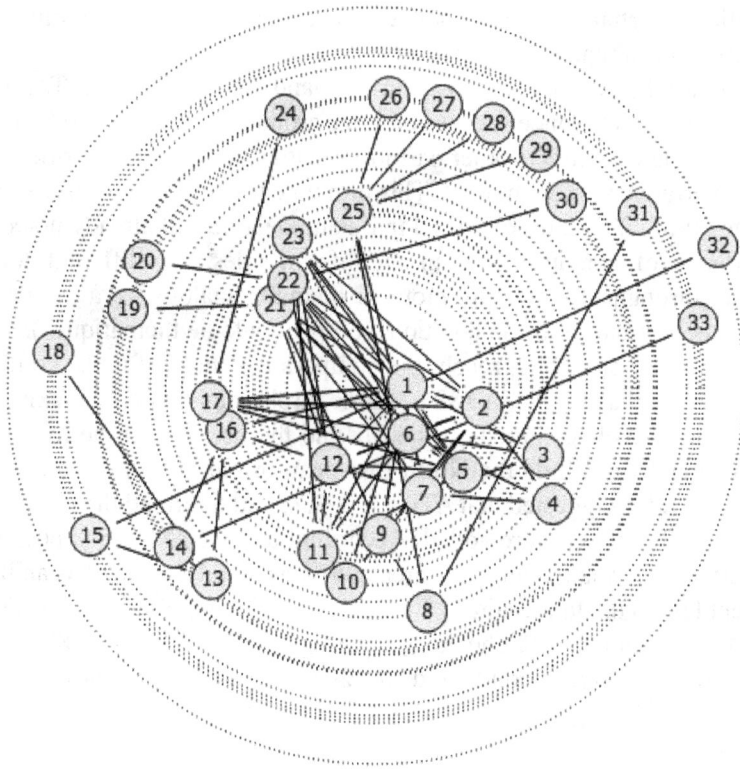

Fig. 9.2.   Closeness centrality calculated from the same corpus.

If the six most central tales are removed, the network fragments into several independent sub-networks, showing that these central tales play a major structuring role. This suggests that the network of tales studied follows the "small-world" model as defined by Watts and Strogatz (1998), a hypothesis I later corroborated using another corpus.

To my knowledge, this is the first time a statistical analysis has demonstrated the attractor role of a set of tales. This central nucleus includes ATUs 1 ("The theft of fish"), 2 ("Tail-Fisher"), 3 ("Simulated Injury"), 4 ("Sick Animal Carries Healthy One"), 5 ("Biting the Tree Root"), and 15 ("Theft of Food by Playing Godfater"), defined by the opposition between the cunning fox and the powerful but unintelligent bear or wolf. Surrounding this central opposition are various narratives, more or less directly drawn to this theme. It does not constitute a complete cycle per se but rather a partial cycle of loosely connected animal tales, illustrating the principle of oral narrative variation (Friedrich, 2014).

One might assume, as Elizabeth Wayland Barber and Paul T. Barber (2004) suggest, that once stories converge around a central theme (e.g., the fox), this nucleus attracts other narratives with significant similarities, creating a narrative attraction point. This principle is not new, as Marie-Louise Tenèze (1958: 297–298) observed that Paul Delarue had already noted the importance of contamination in the life of tales, "at different levels, whether through a connecting motif or thematic analogy." ("à des paliers différents, soit par motif de jonction, soit par analogie thématique").

A more empirical approach, the comparative study of texts, confirms these statistical analyses. In 1888, Kaarle Krohn highlighted the existence of an animal cycle centered on the fox and the bear, which he located in Nordic countries. While the North European origin of this cycle remains debated, Krohn suggested that ATUs 1, 2, 3, 4, and 5 formed the initial core to which other types were added. According to him, ATU 15 appeared in the northern Germanic language area and "was therefore added as an independent tale to the tale chain" ("Es ist also als ein ursprünglich selbständiges Märchen an jene Märchenkette angefügt worden." Krohn, 1888: 80).

Further research conducted in 2019 refined the study of tale networks.

9.3 Fairy Tales

For this new analysis, I again utilized the ATU classification, this time focusing on another category of tales, i.e., "Tales of Magic" (Uther, 2011; ATU 300–794A). This category comprises narratives featuring beings or objects endowed with supernatural powers. The aim of this new study was to explore how these stories interact, identify recurring combinations, and understand how these interactions influence the overall structure of the tale network.

As with the previous corpus, only tale types showing combinations with others were included in the analysis, with each selected type being linked to an average of 4.1 other types. A matrix of connected ATU was created and analyzed using the Social Network Visualizer v.2.3 software (Kalamaras, 2017) based on an adjacency matrix. Several metrics were applied to answer the question: "Which ATU tale type is the most central in the studied network?" These metrics assess the importance or influence of each ATU tale type, each based on distinct assumptions.

9.3.1 *Degree centrality*

DC calculates the number of direct connections a node has — in this case, the number of direct relationships between a tale type and others. A tale type with a high score is often a highly connected node within the network, although some of its connections may involve less significant tale types. The average centrality in this network (Fig. 9.3) is 0.04 (variance = 0.001) with 23 classes. According to this metric, the ATU 313 ("The Magic Flight"), 314 ("Goldener"), 300 ("The Dragon Slayer"), 400 ("The Man on a Quest for His Lost Wife"), and 302 ("The Ogre's (Devil's) Heart in the Egg") emerge as the five most central (see d'Huy 2019 for more information about this and subsequent results).

9.3.2 *Eigenvector Centrality*

Eigenvector Centrality (EVC; Bonacich, 1972) refines DC by assigning a score proportional to the scores of a node's neighbors. This means that connections to influential nodes contribute more to a node's score than connections to less influential ones. This metric highlights the most central tale types among the most interconnected in the network (Fig. 9.4). The tale types ATU 300 ("The Dragon Slayer"), 314 ("Goldener"), 313 ("The Magic Flight"), 302 ("The Ogre's (Devil's) Heart in the Egg"), and 400 ("The Man on a Quest for His Lost Wife") achieve the highest EVC scores.

9.3.3 *Betweenness centrality*

Betweenness centrality (BC) measures a node's role as a bridge within the network, assessing how often it appears on the shortest path between two other nodes. This metric highlights the significance of a tale type in combining and transmitting narrative elements (Fig. 9.5). The tale types ATU 400 ("The Man on a Quest for His Lost Wife"), 313 ("The Magic Flight"), 314 ("Goldener"), 425 ("The Search for the Lost Husband"), and 302

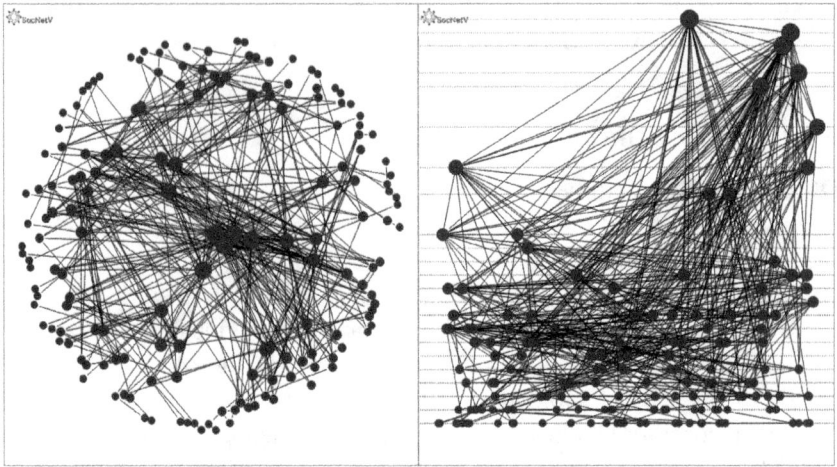

Fig. 9.3. Radial (a) and leveled (b) graphs based on DC. To enhance readability, two graphical representations showing the same results were always used. The meaning of both representations is identical. Correspondence: 1/ 300; 2/ 300A; 3/ 301; 4/ 302; 5/ 302B; 6/ 302C*; 7/ 303; 8/ 303A; 9/ 304; 10/306; 11/ 307; 12/ 310; 13/ 311; 14/ 312; 15/ 312D; 16/ 313; 17/ 314; 18/ 314A; 19/ 314A*; 20/ 315; 21/315A; 22/ 316; 23/ 317; 24/ 318; 25/ 321; 26/ 325; 27/ 326; 28/ 327; 29/ 327A; 30/ 327B; 31/ 327C; 32/327F; 33/ 327G; 34/ 328; 35/ 328*; 36/ 328A*; 37/ 329; 38/ 330; 39/ 331; 40/ 332; 41/ 333; 42/ 334; 43/360; 44/ 361; 45/ 363; 46/ 365; 47/ 400; 48/ 402; 49/ 403; 50/ 403C; 51/ 404; 51/ 407; 53/ 408; 54/ 409; 55/409A; 56/ 412; 57/ 425; 58/ 425A; 59/ 425B; 60/ 425C; 61/ 425D; 62/ 425E; 63/ 431; 64/ 432; 65/ 433B; 66/ 434; 67/ 440; 68/ 441; 69/ 442; 70/ 449; 71/ 450; 72/ 451; 73/ 452B*; 74/ 460A; 75/ 460B; 76/ 461; 77/ 465; 78/ 470; 79/ 470A; 80/ 470B; 81/ 471; 82/ 475; 83/ 480; 84/ 480A; 85/ 501; 86/ 502; 87/ 505; 88/506*; 89/ 507; 90/ 510; 91/ 510A; 92/ 510B; 93/ 511; 94/ 513; 95/ 513A; 96/ 513B; 97/ 516; 98/ 517; 99/518; 100/ 519; 101/ 530; 102/ 530A; 103/ 531; 104/ 532*; 105/ 533; 106/ 537; 107/ 545A; 108/ 550; 109/551; 110/ 552; 111/ 554; 112/ 555; 113/ 556F*; 114/ 559; 115/ 560; 116/ 561; 117/ 562; 118/ 563; 119/564; 120/ 565; 121/ 566; 122/ 567; 123/ 567A; 124/ 569; 125/ 570; 126/ 571; 127/ 571B; 128/ 575; 129/577; 130/ 580; 131/ 590; 132/ 591; 133/ 592; 134/ 610; 135/ 611; 136/ 613; 137/ 650A; 138/ 652; 139/ 655; 140/ 665; 141/ 667; 142/ 670; 143/ 670A; 144/ 671; 145/ 673; 146/ 675; 147/ 681; 148/ 700; 149/ 705A; 150/ 705A*; 151/ 705B; 152/ 706; 153/ 707; 154/ 708; 155/ 709; 156/ 710; 157/ 713; 158/ 715; 159/ 715A; 160/ 720; 161/ 725; 162/ 735; 163/ 735A.

("The Ogre's (Devil's) Heart in the Egg") exhibit the highest BC scores, underlining their pivotal role in the network.

It is important to reiterate that a tale type is an abstract category grouping similar stories, not a tale in itself. This classification is partially arbitrary and may contain misclassifications. Furthermore, the notion of a

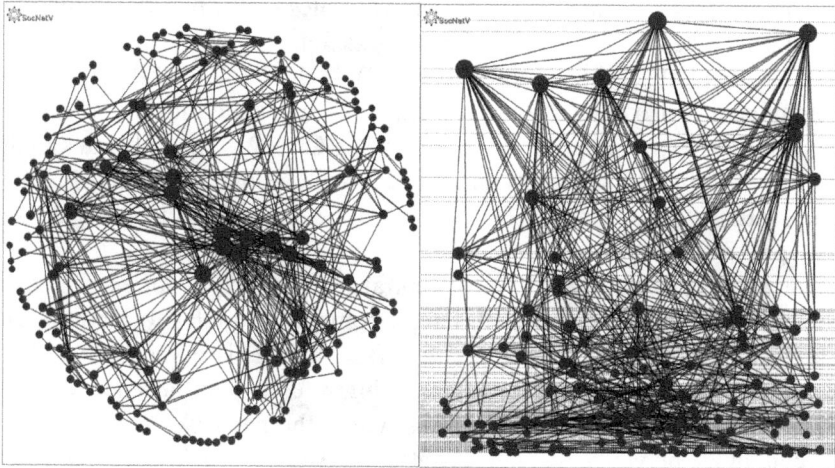

Fig. 9.4. Radial (a) and leveled (b) graphs based on eigenvector centrality.

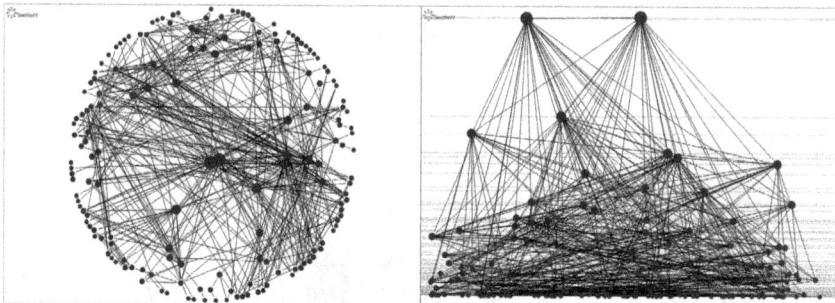

Fig. 9.5. Radial (a) and leveled (b) graphs based on BC.

tale type is broad: it includes both specific narrative stories (e.g., ATU 300) and more general categories encompassing multiple types under a shared structure (e.g., ATU 400). Within this framework, the ATU system is primarily a research tool, valuable for organizing texts and literature.

9.4 "Small World" Networks

The network study of ATU folktales reveals properties similar to "small world" networks described by Watts and Strogatz (1998). It demonstrates a low average distance between tales and strong clustering.

The average distance between two ATU tales — the minimum number of edges needed to connect two nodes — is 3.4.

The local clustering coefficient (CLC) — the likelihood that a node's neighbors are also connected to each other — calculated for the ATU network is 0.67 (variance: DC: 80; EVC: 91; BC: 102). This indicates high transitivity: if A is linked to B and B to C, then C is often linked to A. This score suggests that many ATU tale types are interconnected in a dense network.

Finally, removing the 10 most central ATU tale types for each metric causes the average shortest path length between two tales to increase by 26.2% to 29.9% (DC: 4.4; EVC: 4.4; BC: 4.3). This demonstrates that these highly central tales play a structuring role, as their absence significantly increases the average distances within the network. They serve as crucial links between distinct narrative clusters. Two hypotheses may explain this:

(1) These types act as mental intermediaries for storytellers, linking several sets of folktales;
(2) There are multiple geographical clusters, with combinations of tale types varying by region.

Studying these smaller clusters offers intriguing avenues for future research.

9.4.1 *Influence of "cultural success" and diffusion of ATU types on their centrality*

Could the most connected ATU tales also be the most culturally renowned or expected, such that storytellers sought to include them in as many narratives as possible? For example, about the Seven-Headed Beast (ATU 303, a motif also found in ATU 300), a storyteller explains: "The old folks said there weren't any longer tales than that one [...]. Ah! That story has been listened to by many people! You see, it's the beast attack part that fascinated them." (Félice, 1950: 458; *Les anciens prétendaient qu'on ne pouvait pas dire de contes plus longs que celui-là [...]. Ah! cette histoire-là a été écoutée par bien des gens! Vous comprenez, c'est le coup de l'attaque de la bête qui les intéressait.*)

Ara Norenzayan *et al.* (2006) aimed to identify the culturally most successful tales of the Brothers Grimm. Their study focused on Grimm's tales but included several ATU tale types analyzed here. Using Google searches in English and German that paired each Grimm tale title with the word "Grimm", they estimated each tale's popularity by counting the number of web results.

For this analysis, tales with high cultural success corresponding to specific ATU types were selected. For instance, these included "The Dragon Slayer" (ATU 300), "The Three Stolen Princesses" (ATU 301), and "The Magic Flight (ATU 313)". The centrality scores of these popular tales were then calculated (see d'Huy, 2019 for details).

The assessment of the impact of the geographical distribution of tale types on their centrality scores is complex, due to the lack of specific studies for each type. However, the Types of International Folk Tales repertoire provides an approximate estimate of their geographical spread by country, region, or linguistic area. This repertoire does not account for the entirety of known tale types but only those that have not been documented in major monographs (d'Huy *et al.*, 2017). Thus, if all the tales of a population have been cataloged in a comprehensive monograph dedicated to that tale, and no new version has been collected in writing since then, the population possessing this tale is not explicitly mentioned. Instead, its versions are integrated into the cited monograph to which the repertoire refers.

The following results therefore assume that biases in identifying narrative areas are evenly distributed among the tales, which limits the precision of the calculations.

Although this repertoire does not measure the popularity of a tale in a given region, a correlation has been established between the centrality scores of tale types (including those with high scores, as determined in this paper, as well as those identified as experiencing "cultural success" according to Norenzayan *et al.*, 2006) and the number of cultural zones mentioned in Uther's work (2011) where each tale type is present. The results show a strong correlation between a tale's geographical spread and its centrality score (Table 9.1).

The study of the influence of cultural success and the distribution area on the centrality of ATUs shows that only geographically widespread ATUs, not necessarily the most famous ones, achieve high centrality. Thus, none of the most central tales in this study appear in the list by Norenzayan *et al.* (2006), refuting the idea that their centrality stems from

Table 9.1. Correlation between each centrality measure of a set of ATU (i.e., all tale types with a high centrality score and "most successful" stories identified by Norenzayan *et al.*, 2006: 300, 301, 302, 303, 310, 313, 314, 327, 333, 400, 425, 425C, 440, 450, 461, 480, 510A, 531, 533, 550, 554, 555, 707, 709) and the number of areas identified by Uther.

	DC'	EVC'	BC'
Pearson	0.83 (p = 4.2E-07)	0.80 (p = 2.7 E-06)	0.67 (p = 0.00035)
Spearman	0.87 (p = 3.87E-08)	0.82 (p = 7.06E-07)	0.77 (p = 1.05E-05)

Note: DC', EVC' and BC' are standardized index (0 ≤ index ≤ 1).

a popularity sought by the public. Some well-known stories, such as "Little Red Riding Hood", exhibit low centrality.

The observed correlation, which explains a significant part of data variability (between 45% and 75%), suggests that the most central tale types are also the most geographically widespread and confirm that only geographical diffusion is a key factor in centrality. This wide diffusion appears to be more related to the antiquity of these tales than to their current cultural success. If we accept that the most widespread ATU types are older than isolated ones, then central ATU could date back to ancient periods, possibly even prehistory. This relationship between centrality, diffusion, and antiquity seems to form a relevant explanatory hypothesis.

Let us consider the ATU types present among the 10 most central for each of the measures adopted.

ATU 300, "The Dragon-Slayer", is associated with motifs of fight against a dragon or a serpent, which are found in numerous cultures worldwide. Researchers such as Yuri Berezkin (2014) and Joseph Fontenrose (1980) have demonstrated that this motif is extremely ancient, with Paleolithic roots. Other studies, including those by Kurt Ranke (1934), Michael Witzel (2008), or Julien d'Huy (2023), further support the antiquity of this narrative motif, which is also deeply rooted in Indo-European traditions (Ivanov and Toporov, 1970).

ATU 302, "The Ogre's (Devil's) Heart in the Egg", revolves around a similar motif of vulnerability hidden in an external object, an idea also widely spread across the globe. James George Frazer (1913) studied the worldwide distribution of this motif, linking it to ancestral beliefs.

ATU 313, "The Magic Flight", is also considered very ancient, with evidence of its diffusion dating back to pre-Columbian times, according to research by Franz Boas (1914) and Gudmund Hatt (1949). This motif, which describes the hero's escape using magical means, is absent in Australia and Melanesia. This suggests a Paleolithic diffusion that occurred after the colonization of these regions but before the settlement of the Americas (Berezkin, 2013: 163–167).

Finally, ATU 400, "The Man on a Quest for His Lost Wife", particularly in the version "The Swan Maiden", has been analyzed by various researchers, including Berezkin (2010) and d'Huy (2016), who conclude that the tale likely originated during the Paleolithic era in regions of Southeast Asia and East Asia (see additional references in d'Huy, 2023: 312–315).

To my knowledge, no study has yet been conducted to determine the age and origin of ATU 554, "The Grateful Animals". The fact that this tale is attested in Mesoamerica (Peñalosa, 1996: 66) and widely distributed in North America (Thompson, 1929: Note 46), like ATU 314, could indicate great antiquity. However, this hypothesis remains uncertain: it could just as easily involve independent inventions or later transmissions.

Regarding ATU 531, "The Clever Horse", assessing its antiquity is also complex, as it is a heterogeneous type encompassing several narratives featuring a cunning horse.

These examples show that many of the most central tales in the ATU network could have very ancient origins, possibly dating back to the Upper Paleolithic. This hypothesis is supported by the strong correlation observed between geographical diffusion and centrality in the network, suggesting that the most geographically widespread tales are also the oldest.

These findings, although not exhaustive, help uncover the antiquity of certain narratives that persist in oral traditions and constitute significant "nodes" in the ATU network.

9.5 Conclusion

The study of the ATU network offers new perspectives on the evolution of popular tales, complementing the comparativist mythologist's toolkit and enabling the exploration at a micro level of what phylomythology and areology address at a macro level.

Measures of centrality highlight types that play a pivotal role in the global narrative structure, acting as bridges between different ATU clusters. These central tales are often widely geographically disseminated stories, though not necessarily the most popular in contemporary culture. Their influence appears more linked to their relative antiquity. Thus, this method complements structural approaches (Sergent, 2009), areal studies (Berezkin, 2013, 2017; Le Quellec, 2021, 2022; Thuillard, 2021; Thuillard *et al.*, 2018), and phylo-mythological studies (d'Huy, 2012; 2023b, 2023c), which also provide tools for reconstructing ancient forms of traditions and oral tales.

Finally, the analysis also shows that the removal of the most central tales significantly fragments the network, increasing the average distance between the remaining tales. This highlights the importance of these narratives in transmitting and combining narrative motifs across cultures, as well as the existence of geographic clusters where certain tales tend to combine particularly strongly.

References

Barber, E. W., & Barber, P. T. (2012). *When They Severed Earth from Sky: How the Human Mind Shapes Myth* (Princeton University Press).

Belmont, N. (2013). Manipulation et falsification des contes traditionnels par les cultures lettrées, *Ethnographiques.org*, 26 (in French).

Berezkin, Y. (2010). Sky-maiden and world mythology, *Iris* 31 (UGA Éditions), pp. 27–39.

Berezkin, Y. (2017). Roždenie zvezdnogo neba: Predstavlenija o nočnyh svetilah v istoričeskojdinamike. Saint Petersburg: MAÈ RAN (in Russian).

Berezkin, Y. E. (2014). Serpent that closed sources of water and Serpent that devours nestlings of giant bird: Assessment of the age of the dragon-fighting myths in Eurasia, *Aramazd: Armenian Journal of Near Eastern Studies*, 8(1–2), pp. 178–185.

Boas, F. (1914). Mythology and folk-tales of the North American Indians, *Journal of American Folklore*, 27(106), pp. 374–410.

de Felice, A. (1950). Contes traditionnels des vanniers de Mayun (Loire-Inférieure), *Nouvelle Revue des Traditions Populaires*, 2(5), pp. 442–466 (in French).

d'Huy, J. (2012b). Le conte-type de Polyphème: Essai de reconstition phylogénétique, *Mythologie française*, 248, pp. 47–59 (in French).

d'Huy, J. (2018). Le Web du Goupil, *Mythologie française*, 271, pp. 15–19 (in French).

d'Huy, J. (2023a). Aux origines du dragon: réévaluation d'un article de 2016, *Nouvelle Mythologie Comparée/New Comparative Mythology 2019–2020*, 7 (in French).

d'Huy, J. (2023b). *Cosmogonies: la Préhistoire des Mythes* (Paris: La Découverte; in French).

d'Huy, J. (2023c). *L'Aube des Mythes* (Paris: La Découverte; in French).

d'Huy, J., Le Quellec, J.-L., Berezkin, Y., Lajoye, P., & Uther, H.-J. (2017). Studying folktale diffusion needs unbiased dataset, *Proceedings of the National Academy of Sciences*, 114(41), E8555. Available at: https://doi.org/10.1073/pnas.1714884114

d'Huy, J., Le Quellec, J.-L., Thuillard, M., Berezkin, Y. E., Lajoye, P., & Oda, J. (2023). Little statisticians in the forest of tales: Towards a new comparative mythology, *Fabula*, 64(1–2), pp. 44–63.

Fischer, J. L., & Lüthi, M. (1977). Assoziation. K. Ranke, H. Bausinger, W. Brückner, M. Lüthi, L. Röhrich, & R. Schenda (eds.), *Enzyklopädie des Märchens: Handwörterbuch zur historischen und vergleichenden Erzählforschung. Band 1* (in German).

Fontenrose, J. E. (1980). *Python: A Study of Delphic Myth and Its Origins* (Berkeley, USA).

Frazer, J. G. (1913). *The Golden Bough: A Study in Magic and Religion* (London, Royaume-Uni: Macmillan).

Friedrich, U. (2014). Zyklus. K. Ranke, R. W. Brednich, & Akademie der Wissenschaften (eds.), *Enzyklopädie des Märchens: Handwörterbuch zur historischen und vergleichenden Erzählforschung. Band 14, Vergeltung-Zypern, Nachträge: Abi-Zombie (in German)*.

Hatt, G. (1949). *Asiatic Influences in American Folklore*. Kobenhavn: I kommission hos Ejnar Munksgaard.

Ivanov, V., & Toporov, V. N. (1970). Le mythe indo-européen du dieu de l'orage poursuivant le serpent: reconstruction du schéma, P. Maranda & J. Pouillon (eds.), *Échanges et communications: Mélanges offerts à Claude Lévi-Strauss, à l'occasion de son 60e anniversaire*, pp. 1180–1206 (in French).

Kalamaras, D. (2014). Social Network Visualizer (SocNetV), *Social Network Analysis and Visualization Software*. Available at: http://socnetv.org.

Karsdorp, F., & Van Den Bosch, A. (2016). The structure and evolution of story networks, *Royal Society Open Science*, 3(6), 160071. Available at: https://doi.org/10.1098/rsos.160071.

Kawan, C. S. (1996). Kontamination, K. Ranke, R. W. Brednich, & Akademie der Wissenschaften, (eds.), *Enzyklopädie des Märchens: Handwörterbuch zur historischen und vergleichenden Erzählforschung. Band 8*, pp. 210–217 (in German).

Korotayev, A., & Khaltourina, D. (2010). *Mify i geny: Glubokaja istoričeskaja rekonstrukcija*. Moscou, Fédération de Russie: Librokom (in Russian).

Krohn, K. (1888). *Wolf Bär und Fuchs: eine nordische Tiermärchenkette*. (O. Hackman, Trans.). Helsingfors, (Finlande: Finn. Litteratur Gesellschaft; in German).

Le Quellec, J.-L. (2021). *Avant nous le Déluge! L'humanité et ses mythes*, (Bordeaux: Editions du Détour; in French).
Le Quellec, J.-L. (2022). *La caverne originelle: Art, mythes et premières humanités*, (Paris, France: La Découverte).
Le Quellec, J.-L., & Sergent, B. (2017). *Dictionnaire critique de mythologie*, (Paris: CNRS éditions; in French).
Luzel, F.-M. (1887). *Contes populaires de Basse-Bretagne*, (Paris, France: Maisonneuve et C. Leclerc; in French).
Mac Carron, P., & Kenna, R. (2012). Universal properties of mythological networks, *Europhysics Letters*, 99(2), Available at: https://doi.org/10.1209/0295-5075/99/28002.
Mauss, M. (1903). Les mythes. E. Durkheim (ed.), *L'Année sociologique* (Paris: Félix Alcan; in French), pp. 243–240 (in French).
Norenzayan, A., Atran, S., Faulkner, J., & Schaller, M. (2006). Memory and mystery: The cultural selection of minimally counterintuitive narratives, *Cognitive Science*, 30(3), pp. 531–553. Available at: https://doi.org/10.1207/s15516709cog0000_68.
Peñalosa, F. (1996). *El cuento popular maya: una introducción*. (Rancho Palos Verdes: Ediciones Yax Te'; in Spanish).
Propp, V. J. (1928). *Morfologija skazki* (Leningrad, Russie: Academia; in Russian).
Proulx, S. R., Promislow, D. E. L., & Phillips, P. C. (2005). Network thinking in ecology and evolution, *Trends in Ecology & Evolution*, 20(6), pp. 345–353.
Ranke, K. (1934). *Die zwei Brüder: eine Studie zur Vergleichenden Märchenforschung. FF communications* 114 (Helsinki, Finlande: Suomalainen Tiedeakatemia, Academia Scientiarum Fennica; in German).
Röhrich, L. (1976). *Sage und Märchen: Erzählforschung heute* (in German).
Sergent, B. (2009). *Jean de l'ours, Gargantua et le dénicheur d'oiseaux* (La Bégude de Mazenc, France: Arma Artis; in French).
Ténèze, M.-L. (1958). Une contribution fondamentale à l'étude du folklore français: le conte populaire français. Catalogue raisonné des versions de France et des pays de langue française d'outre-mer, Tome I, *Arts et Traditions populaires*, 6(3-4), pp. 289–303 (in French).
Thompson, S. (ed.). (1929). *Tales of the North American Indians* (Cambridge: Harvard University Press).
Thuillard, M. (2021). Analysis of the Worldwide Distribution of the "Man or Animal in the Moon" Motifs, *Folklore: Electronic Journal of Folklore*, 84, pp. 127–144. Available at: https://doi.org/10.7592/FEJF2021.84.thuillard.
Thuillard, M., Le Quellec, J.-L., d'Huy, J., & Berezkin, Y. (2018). A large-scale study of world myths, *Trames: Journal of the Humanities and Social Sciences*, 22(4), p. 407. Available at: https://doi.org/10.3176/tr.2018.4.05.

Uther, H.-J. (2011a). The types of international folktales: a classification and bibliography: Based on the system of Antti Aarne and Stith Thompson. *Part I, Animal Tales, Tales of Magic, Religious Tales, and Realistic Tales, with an Introduction.* (Vol. 1) *FF Communications* 284. (Helsinki, Finlande: Suomalainen Tiedeakatemia).

Voigt, V. (1977). Affinität. K. Ranke, H. Bausinger, W. Brückner, M. Lüthi, L. Röhrich, & R. Schenda (eds.), *Enzyklopädie des Märchens: Handwörterbuch zur historischen und vergleichenden Erzählforschung. Band 1* (in German).

Watts, D. J., & Strogatz, S. H. (1988). Collective dynamics of 'small-world' networks, *Nature*, 393(6684), pp. 440–442.

Witzel, E. J. M. (2008). Slaying the dragon across Eurasia. J. D. Bengtson (ed.), *Hot Pursuit of Language in Prehistory: Essays in the Four Fields of Anthropology: In Honor of Harold Crane Fleming* (Amsterdam, Pays-Bas: John Benjamins Publishing Company), pp. 263–286.

Index